THE GARDEN IN
WINTER

THE GARDEN IN
WINTER

ROSEMARY VEREY

TIMBER PRESS • PORTLAND, OREGON

This book is for my gardening friends on both sides of
the Atlantic who love their gardens in winter, even though
the ground is snow-covered and frozen; those who appreciate the beauty
of crisp, frosty days and of sweet-smelling winter flowers;
those who like to be outside on sunny days, because
they know that their gardens will only be beautiful
in summer if the winter work is done.

I have talked to many old friends and made many new ones while
writing this book. I dedicate it to them all.

© Frances Lincoln Limited 1988
Text © Rosemary Verey 1988

The Garden in Winter was edited and designed
by Frances Lincoln Limited,
4 Torriano Mews, Torriano Avenue, London NW5 2RZ

Paperback edition first published in North America in 1995
by Timber Press, Inc.
The Haseltine Building
133 S.W. Second Avenue, Suite 450
Portland, Oregon 97204, U.S.A.
1-800-327-5680 (U.S.A. and Canada only)

ISBN 0-88192-337-0

Printed and bound in Hong Kong by Kwong Fat Offset Printing Co. Ltd

CONTENTS

A PLEA FOR THE GARDEN IN WINTER

A GARDEN IN WINTER is the absolute test of the true gardener. Fair-weather gardeners are to gardens what interior decorators are to buildings – they know only half the story. True gardening is as much about the bones of a garden as its planting; true architecture is as much about the form and structure of a building as its rooms. If your garden looks good in winter, you belong to a select band capable of bending nature to its will. If you master the worst that winter can throw at you – ice, snow, wind without remorse – you will have a sense of conquest.

Anne Scott-James confessed in her book *Down to Earth* to being a latter-day convert to winter gardening: 'Dilettante gardeners love the spring and summer, real gardeners also love the winter and I've joined the club... I've grown to love the garden in winter.'

I too have grown to love my garden in winter. Yet, many years ago, when I began to think about it, one thing puzzled me. Why did the countryside look so attractive and my garden so dead? It struck me then that a strange reversal takes place in the fortunes of countryside and garden from summer to winter. In summer our gardens are *en fête*, with their bright swathes of colour, while the countryside is demure in shades of green with the occasional pocket of wild flowers. In winter it is the countryside that looks marvellously rich, with its rolling browns and greens and hedgerows popping with berries, while gardens may well be bare until March signals the start of spring.

Winter lays bare the simple structure of this small, Colonial-style garden in Princeton, New Jersey. In the summer it is leafy and inward-looking, with ivy in the central bed trailing over the brick paths that lead to the garden's focal point. In winter the design opens out, revealing the tracery of the deciduous trees beyond. Well-defined beds, edged in boxwood, are emphasized by a mantle of snow.

I determined to bring life to my garden in winter – to make autumn join hands with spring. Winter was to be a season in its own right, vital to the gardener who really wants to garden. I decided, like that innovative gardener, William Robinson, to banish the idea that 'winter is a doleful time for gardens'.

First of all I had to remember the dictum of the great gardener and garden writer, Christopher Lloyd: 'A year without its winter would seem all wrong to me, as a countryman. It is a necessity, not just a necessary evil... To think of leaf fall and the onset of winter as sad is anthropomorphic: the usual silly way we have of looking at things from our own point of view.'

I had to shed my preconceptions about all sorts of things – about the weather and the time of day, about working in the garden for odd moments, about the plants I knew I liked and those I thought I didn't. I had to begin to appreciate winter scents, to notice the colour and texture of tree bark, to discover lichen and moss on the walls, to watch winter buds open and bulbs push their way through the soil – leaves first or sometimes flowers first, as with the winter aconites. I had to learn which berries the birds do not immediately devour and to work out where the sun, low in the sky, made the strongest slanting shadows on the lawn.

I began to realize that the structure of my garden is even more important in winter than in other seasons, because the bones become apparent and the eye is not distracted by beguiling planting. So the framework of my garden had to be set in winter. Paths, walks, hedges, allées, vistas, all would determine its form. I also had to realize that winter's beauty – clear and spare – is quite different from the freshness of spring blossom, the lushness of summer flowers or the richness of autumn leaves.

Other lessons were taught me by the plants themselves. Looking at them afresh, I began to think about those which keep their beauty – of leaf, berry or flower – through the winter, and to discover those whose beauty appears only with the onset of winter. The main barrier I had to break down was mental – the notion that plants are either for winter, or for summer. Looking at each tree or shrub as one thread in a twelve-month tapestry is more demanding but certainly more rewarding than closing the door on the garden in October and not opening it until the following March. I learned to think of planting in three dimensions – to sink plants into the ground in layers, so that bulbs will pierce through groundcover, spreading under shrubs that act as hosts to climbers – to place plants in the garden so that they become foils to each other season by season, colour by colour and texture by texture.

If our gardens are to be more than graves commemorating summer's beauty, we must start by using our eyes. The problem all too often is that, when we look, we do not see. We fail to appreciate to the full the beauty around us. And if we do not look, beauty will be denied to us. 'It is the old story of "eyes and no eyes",' wrote the great Edwardian gardener Vicary Gibbs. 'Given bright sunlight, without which no colours can be fully seen, there they are if we will only observe them, and the more we look the more we see. It is the perfect harmony of Nature's work which hides her brilliant hues from the careless, though to the patient watcher she reveals fresh beauties both of form and colour every day ... things which many a countryman has lived and died without noticing.'

Today, mid-December, the light in the Cotswolds is electrifying, brilliant. The air is cold and crisp, the sky a reflecting blue, the earth black, the leaves and the sky motionless, the clouds a thin haze on the horizon. The only movement to be seen is the swift flight of the finches, tits, sparrows and the occasional wren as they search for food in bushes and borders. Wherever I look, there is a feeling of repose and happiness among the plants. The cool temperate plants are, as Christopher Lloyd has pointed out, 'not only well adapted to winter, they need and could not do without this season. You only have to see what a poor account of themselves roses give in the tropics, and how short-lived they are, to realize that without winter, plants such as these burn themselves out. Others need cold to bring about the physiological changes that enable them to break dormancy the following spring. They are lost without it, remaining dormant forevermore.'

Then there are those glorious days when the ground is covered with snow. I admit I love snow particularly for the moments when, with a clear conscience, I can stay indoors and watch from inside – it is beautiful even when the snow is falling thickly and I am seeing the garden through a veil.

The evening before, the clouds will have started to look heavy, and it is no surprise next morning to wake to that familiar white light bouncing off the ceiling, a reflection of the snow carpet outside. Look out and all is clean and pure. If the fall has been light, the shrubs and trees stand out crisply. As the sun appears the shadows have clear, strong outlines. There is a muffled silence; familiar sounds are absorbed, distant noises come closer. For the first few hours no human footprints scuff the surface, but there are the tell-tale marks of a rabbit or a hare or sometimes a fox taking a short cut through the garden in search of breakfast. I try hard to persuade everyone to keep off the unmarked snow.

If the fall has been heavy, the contours of the garden are transformed. Each shrub is quite enveloped, each tree sagging with its unaccustomed burden. This protective blanket safeguards precious plants from sharp attacks of frost and cold; far better that they should stay covered than that the snow should melt too quickly, leaving the sap to freeze in the stems. But this advantage is secondary to the beauty of the garden under the snow. I like to think of Vita Sackville-West planting her White Garden at Sissinghurst 'under the first flakes of snow'.

For me, frost has a strange appeal, the magic of waking to find the garden white with frost, stems hoary-haired and leaves outlined with rime, as though a paintbrush dipped in silver had been drawn assiduously round each one. The heads of the sedums, the ivy leaves, the brown seed heads of *Phlomis fruticosa* have all changed colour to silver or grey. Everything has taken on a light and fragile air. On days like these, while the grass is still frozen and the ground hard, you must not walk on your lawn. If you do, your footsteps will stay, first as imprints in the frost and later as unwanted brown marks. Of course, there are some days when the frost is deadly – that awful frost after which we cannot know for days or even weeks whether a shrub has succumbed or not. If it does, it will invariably be one we felt we could not afford to lose.

Even the most ardent admirer of winter has to reckon with those rain-soaked days when braving the elements would seem like a nightmare. But many years ago I learned that if you wake up on a winter morning with the rain beating against your window, although you may feel there is little incentive to go outside, the reality is quite different. Get on a horse or take a walk along a country lane and somehow the wind is never so fierce nor the rain so unpleasant as you

imagined. You will see pheasants in their winter plumage, hares crouching in the stubble, old man's beard clinging to the hedgerows, hips and sloes amidst the skeletons of thistles and cow-parsley. When you get back home, your cheeks will be glowing and your approach to the day positive.

Some of us are forced by the elements into days, if not weeks, of armchair gardening. Even if we could venture out of doors, we need to to curl up in front of the fire, read our newest or most treasured gardening books, browse through catalogues, and get out squared paper to make new designs. When the door is shut against intruders and the curtains drawn against the distractions of the view outside, you can think of your garden in the abstract and draw together the threads of summer and winter planting.

There are plenty of other gardening jobs you can tackle

A yew arch in the making. Rime and winter sunlight bring out the colour contrast between the dark yew hedges in the foreground and the silvery expanse of frost-laden grass beyond, marking a dramatic transition from a formal garden – complete with statues – to a natural meadow planted with fruit trees.

inside – planning your vegetable programme, scrubbing flowerpots, writing labels. And if the trees are swaying and leaning, almost shouting out to you from which direction they are being buffeted, this is the day to plan your windbreaks.

To teach yourself to 'see' your own garden in winter, look first at the trees and shrubs planted for their spring or summer flowers, their handsome leaves, their autumn colours. In their winter guise they will have a different allure. Each has its own winter character, with buds of varying shape, size, colour and texture. The winter buds of the birches are silver-brown, thin and pointed, those of *Sorbus sargentiana* red, fat and sticky, and those of *S.* 'Embley' dark red and sharp as needles. The ash buds are black at the ends of their upward-turning twigs. The leaf buds on *Paeonia delavayi* are deep brown – if they start to look black I am nervous that they may be dead. By Christmas the buds on the lilac bushes are a surprising green. The flower buds on *Parrotia persica* are small and jet black; then in February they suddenly split open to reveal a tiny scarlet flower. The felted buds of *Magnolia* x *soulangiana* 'Lennei', which are there throughout the winter months, sometimes open to reveal a few precocious flowers. The one thing they all have in common is that they are waiting for spring's warmer days. Each year the point at which they change colour or open will differ – not dramatically so, but with enough interesting variations to keep you looking, observing, discovering.

Winter brings unexpected colour, too, when the frost bites. The mahonias we grow for winter scent and year-long architectural form are a dramatic example: some of their spiny leaves turn fiery red when the cold strikes them. Even the scentless common *Mahonia aquifolium* should not be scorned. It is worth admiring for its shiny dark leaves which turn ruby-red in winter.

The low-slanting winter sun plays new tricks with the light. Have you noticed how it catches the leaves on the evergreen trees? All through the winter months the sun's rays highlight the sides of trees and bushes; many of the golden leaves glow more than they do in summer. My four 'Golden King' hollies near the house, the golden privet in the distance and the euonymus climbing up the pear tree are all gilded by the shifts of light, the luminosity of their leaves changing as the sun moves from east to west.

In winter, rainy days enrich the colour of even quite ordinary things in my garden. When water lies in puddles on irregular paving, sodden dark brown leaves attract the eye. On the acanthus the deep green leaves glisten. The garden seats become deeper grey; when they dry they are silvery in the sunlight. Saturated with rain, my stone walls change colour too, their lichen and moss darker and more prominent.

Winter invests plants with an extraordinary intensity of colour. Each year, halfway through winter, I suddenly see orange jewels on the wet path – the seed pods of *Iris foetidissima,* newly cracked open, spilling and shooting their berries everywhere – or has a mouse carried them off? This year for the first time I noticed, hidden behind the leaves of the yellow euonymus, brilliant orange berries with vermilion centres, and the unusual winged stems of *Euonymus alatus.*

Even on dull, sunless days, when the clouds are heavy and the light flat, the pale pink berries of *Sorbus hupehensis* look like blossoms against the dark grey sky, and the pale green flowers of *Helleborus foetidus* and *Euphorbia characias* appear sprayed with luminous paint. The hanging flower heads of the euphorbia, tucked away near a hedge for protection, briefly display their conspicuous black spots surrounded by green; it is always sad when the flowers on this, the 'frog's spawn bush', are caught by the frost before they have had time to open. In my vegetable garden the sight of the leaves of the Verona chicory and of the purple cabbages makes picking the Brussels sprouts a joy instead of a chore.

The downy seed heads of the autumn-flowering clematis, especially *Clematis tangutica,* look as though they are made of the finest feathers when they are dry and the sun shines through them. Perfect in their symmetry, they wait to take off in a strong wind to distribute their seeds far and wide. Pick a handful and they will remain unchanged in a vase forever. In contrast, by December the phlomis seed heads become quite hard and their stalks brown and brittle. The ivy drupes too undergo a startling metamorphosis. One week they are brown and their carpels fused; only their perfect globular shape seems worthy of notice. Then, as they open, all of a sudden they show streaks of green one day; the very next day each head is green all over, except for the five brown carpels, now split on the periphery.

The special quality of winter light catches the feathery seed heads of autumn-flowering Clematis tangutica *and highlights the orange-scarlet hips of* Rosa rugosa *'Rubra'. These plants provide a more subtle association of colours and textures in winter than do their flowers and leaves in summer. Brightly coloured hips come as a surprise among the gentler shades of green, grey, brown and yellow.*

For scent and colour, Viburnum x bodnantense *is one of the best winter-flowering shrubs, carrying double clusters of scented, rose pink blooms along its leafless stems from December right through to February. Hardy and frost-resistant, it can be relied upon to provide a welcome waft of scent and splash of unexpected colour during winter's starkest months.*

Noticing each change as it occurs is a key element in learning how to make your garden attractive in winter. Another is actually to record, season by season, variations in the time things happen. After a few years you will find you know your garden – and, if you are a countryman, the lane and fields surrounding it – as well as you know your best friend. Keeping a diary is a wonderful discipline. Some of the great country diarists – the Rev Francis Kilvert and Gilbert White – have become household names. The 18th-century gardeners' *Kalendars* have snippets of practical wisdom that are as relevant now as then.

One of the rewards of watching changes in the garden throughout the winter is to observe at first hand the balance nature strikes between plant and animal life. Seed- and fruit-eating birds cannot survive without their natural foods; plants regenerate more freely after their seeds have been digested and excreted by birds, often quite a distance from the parent plant. Each benefits the other.

A garden without birds would be dreary indeed. In summer I enjoy them for their song and their young. In winter I love their quick flight across the garden, their forays among the hips, their sudden bursts of song when the sun shines even briefly. Most enjoyable of all is their appreciation of the food we give them to eat. When nature's provisions become scarce, locked in ice and snow, then the birds flock to the bird table. Suddenly the first tits appear, or the house sparrows. On really cold days, hungry, impatient tits actually peck on the glass panes, demanding their due. It is exciting too when the unexpected visitors arrive – the green woodpecker which lives close by, or a few fieldfares which have 'happened' on the garden from the hedgerows. Sometimes it is a bird I do not recognize, and out comes the bird book.

Another pleasure – a more surprising one perhaps in winter than in summer – is scent. There are always so many summer scents that it is frequently difficult to pinpoint any one in particular. In winter, however, they come not in battalions but in single file. All the more refreshing because of their scarcity, they are strongest at noon, when there is still warmth in the sun's rays. Because they are few and far between, think carefully about placing those shrubs which have scent to offer in winter. Space them out in your garden so that you can move from one delight to the next. Some – winter honeysuckle and winter sweet – will waft upon the air, while others have to be picked and held in the hand or brought indoors.

We are often encouraged to place fragrant winter shrubs

near the house, so we can enjoy them without having to go too far down the garden. I no longer agree with this notion, though I used to. Perhaps the most important reason for my change of mind is that the majority of them have rather uninteresting, even coarse, leaves. They are the last things you want to have occupying prominent places by doorways or in borders close to the house when summer comes. One maybe, not more. A more positive reason is that we should walk to find them. Who knows what else we may find unexpectedly on the way? I love strolling through my shrub area and coming on a viburnum with an alluring scent. My two favourites for winter scent are *Viburnum farreri* and *V.* x *bodnantense* – wonderful shrubs for the wild garden and also for the shrub border, where you can use them as host plants for summer-flowering clematis.

There is another, more commonsense, reason for spreading your winter scents around the garden. Part of the joy of summer-scented plants is to have their fragrance wafting into the house through the open windows, but who wants cold air rushing in for the sake of scent in winter? Better to pick your branches, bring them indoors and leave the windows shut.

Evergreens have their own individual aroma. Crush a leaf of the western red cedar and I guarantee you will be reminded of parsley. Yew has its own, almost mysterious, scent. Walk through a pine wood, shut your eyes and you could be in the now-familiar pine bath. Brush against a rosemary bush on a dry day to release the scent; on a rainy day *Helichrysum italicum* (syn. *H. angustifolium*) will remind you immediately of curry.

Quite apart from the scent of plants, winter has its own distinctive smells. Fog, for instance – those mornings when the air is heavy, thick and damp, a damp more pervasive even than rain. But what I love best is the smell of the soil, a rich brown soil, well manured or covered with leaf mould. On the east coast of America, it is the smell of salt hay that is distinctive. Winter's smells are now fresh and light, now dank and sweet, but one we all remember from childhood days is the smoky, woody smell of the bonfire. The struggle to start a winter bonfire and to keep it smouldering involves an endless chain of stoking, raking and compacting. As the smoke blows, first one way, then the other, our nostrils fill with that unforgettable acrid tang.

One of my principal challenges in winter is to find something in the garden to decorate my house. At the start of winter, the task is made easier by the way autumn extends her hand with gifts of berries, seed heads and the last remaining blooms on the roses, the penstemons, the asters, the liriopes and the wonderful pansies and violets. But when frost has finally withered these flowers, what then? Nothing until spring? There is such a pleasure to be gained from going out and picking a bunch of flowers, however humble, from your own garden, especially at a time when they are scarce in the countryside and other people's gardens. Vita Sackville-West concurred. 'How precious', she wrote, 'are the flowers of mid-winter – the genuine toughs that for some reason elect to display themselves out-of-doors at this time of year.' When I thought about my own band of toughs, I was surprised how many there were. Because a single flower in winter is worth any number in summer, an essential part of my plea is to ask you to plant as many of them as you can in your own garden.

The garden is bound to vary from winter to winter, from country to country. I am well aware that in many American states, winter is much longer and much colder than ours – the snow deeper, the moment of thaw more dramatic and the onset of spring swifter. I appreciate that my task of nursing my garden through winter is much easier than that of my gardening friends in the Midwest – I am glad to be in England. But then if God had standardized the weather, half the joy of gardening would be lost.

SPACE, STRUCTURE & PATTERN

*I*F YOU ARE fortunate enough to walk in a garden which looks and feels good in mid-winter, you will realize that it does so because of its use of space, the patterns created by its paths and walls, the shapes of its shrubs, the shadows of its evergreens and the silhouettes of its tree trunks and twisted branches. This is the sort of garden you enjoy looking at from your window on winter days and in which you will get satisfaction from working outside, watching the buds on the shrubs swell and open. We hope the weather will allow us to do both.

Planning the garden to take account of winter requires an overall framework and ground plan, as well as a planting scheme. The framework is provided by the vertical elements – hedges and walls and fastigiate trees – and the ground plan by paths, border shapes and lawns. These are the permanent features that remain more or less the same each season of the year, but whose character is more apparent during the winter, when colour distracts less. This is the structure of the garden, and it must be considered for its effect in summer as well.

My thoughts are mirrored by those of Sir Roy Strong, until recently the director of the Victoria & Albert Museum in London, for whom gardening is a passionate hobby: 'What fascinates me about the garden in winter is that it is the true test of garden design, for without foliage and shrubs a garden stands or falls for its compositional value on clipped evergreens, evergreen trees and shrubs and the exact placing of

A winter view of the Canneman garden in Neerlangbroek, the Netherlands, reveals the detail of its all-seasons framework. A heavy fall of snow accentuates the symmetric arrangement of the evergreen hedges which line the axial pathways and enclose each outdoor room. Promising an abundance of flower and foliage colour in summer months, upright and deciduous shrubs provide the perfect foil to a disciplined layout.

statuary, urns and gazebos. It reminds one that gardening, as it emerged in the grand manner during the Renaissance, did not depend on flowers for its effect. In a way the garden becomes like a church in Lent, stripped of its furnishings, so that one's eye falls back on to the purity of the architecture, devoid of the distraction of ornament.'

The way space in a garden is divided is the fundamental basis of its design. Will you divide yours up with solid partitions such as walls and hedges, or with a trellis, pleached trees, a group of shrubs or a stone colonnade? Pleached trees are hedges on stilts, with a pattern of trunks between which you can see and walk; a colonnade is another such division. Walls and hedges completely cut off views beyond, except through their exits and entrances; other divisions will hold you in suspense or allow glimpses beyond. You must decide.

A first impression often stays with you when others fade. I have visited many gardens since we began reshaping ours in 1960, but the impact of one garden in particular remains with me. I have remembered all the lessons it taught me.

In this one small garden was a room, a space, an allée and an orchard, all fitting into an area perhaps not more than 18m (60ft) square. It was a miniature Hidcote or Sissinghurst – a narrow strip, flanked by hedges opening out into a bulge at the end. Each element was an entity in its own right, each had its own character, leading you gently from one mood to the next. I visited it in high summer, but the design and planting were such that a walk through it in winter would be an enjoyable experience – the first space with evergreens and patterns of box edging, the next made interesting by pretty paving and probably some herbs, the allée with shining green hollies, and the orchard with well-pruned fruit trees. Small though it was, this garden felt spacious and was full of interest.

From its front garden, which was invitingly well planted, you walked around to the side of the cottage. First there was

Above: A wrought-iron gate set into a dense evergreen hedge allows a glimpse beyond the boundary of this garden under snow.

Right: Winter extends summer boundaries and broadens vistas. From across the knot garden to the side of the house at Barnsley, the Irish yews stand out starkly in the distance, while in the foreground clipped balls of Golden King hollies provide winter colour and form.

a tiny formal area, with box-edged beds full of summer flowers. I cannot possibly remember now what they were – it was the general impression which struck me. There were climbers up the cottage wall, and evergreen shrubs and hedges planted to hide the garden from the village street. Everything was in scale – the dwarf box, the small beds in keeping with the size of the cottage. There were narrow paved paths between the beds, and a seat to act as a focal point as well as a place for sitting out in the evening. A rustic stone archway in the centre of a rose hedge led into the next compartment, this time an informal space. It was quite different in character – filled to overflowing with cottage garden flowers, a path winding its way, serpentine fashion, through random beds. The infilling, not the shape, was important here. In winter it would have been a place to walk through from the formal garden, away from the house, under another archway, and into an impressive cross axis – tall hollies on one side and a line of standard hawthorns clipped into balls on the other. This was the boundary of the garden. Turning right past the hollies, another defined space opened up. Here were fruit and rose trees growing through meadow grass, enclosed by a hedge or fence with informal planting. And so back to the house.

My garden is designed less like a series of outdoor rooms, more like an open-plan house. It is surrounded by walls and hedges, but the divisions between each space are low parterre borders with high shrubs, open railings and a pleached lime walk. You have the feeling of moving from one area to another, without the hindrance of solid boundaries. When the shrubs have lost their leaves, the plan is even more open – the vistas becoming broader and the overall view more extensive. The evergreens stand out more starkly – I am thinking particularly of the yews which march along my central vista leading from the house. Beyond the yews the white trunk of a silver birch provides the perfect punctuation mark. Through the fine tracery of its branches the view appears to go on and on. The railings, covered with climbers in summer, become in winter the definite division between the pond garden and the parterre. Surprise vistas are provided by the classical temple in one direction and the lime walk and laburnum avenue in the other. The wild garden is tucked away across a lawn.

Roy Strong's garden, created out of a field in Herefordshire, developed over a twelve-year period. A series of photographs taken each year records the progress of its hedges and topiary – for this is what the garden is all about. With the emphasis on geometry and architecture, its design is based on 16th- cen-

tury gardens in which hard surfaces, steps, terraces, statues and urns, as well as clipped hedges and topiary, were important structural elements. Much of the garden was drawn first on paper but Roy Strong designed his Serpentine walk by outlining it in white paint on the rough grass.

Wholly formal, at every turn vistas are revealed off the main allée and fresh surprises await you as you peep through a gap in the hedge into a small secret garden. The whole place retains an air of mystery, conferred by its hidden enclosures. Different emotions crowd in as you walk from one enclosure to the next, along an allée or towards a vista, on the flat or uphill.

Moving into the most important allée – its hedges 1.8 m (6 ft) apart, its garden rooms opening out along it, its descending steps punctuated by carefully placed urns, and even its yew hedges clipped into 'steps' – I realized how the dark yews gave it serenity while emphasizing its drama. It was the abrupt ending of the view which came as such a surprise. There should have been the gates into paradise; instead was a comfortable seat on which to sit and think. In 1987 the hedge was cut through to extend the vista and to make a further descent into a new garden with contrasting, informal, naturalistic planting, the needed paradise.

The hedges in this garden are mostly yew, sometimes beech, occasionally laurel and cypress, all easily clipped and shaped, and all evergreen except the beech, which has a wonderful rustle to the tawny leaves on windy winter days.

Left: Even in winter, the vistas in this formal garden are contained within boundaries. Like a continuous stage set, hedges create a series of outdoor rooms, each providing the setting for a different scene. Framed by rich brown beech, the Jubilee garden is enclosed with common laurel interspersed with conifers. A brick path crosses the garden, a broken stone circle in the centre featuring an armillary sundial. In the spring and summer the colour theme ranges from white flecked with shades of violet to purple but the garden is planned to look and feel good all year round.

Above: Evergreen hedges provide a permanent backdrop to carefully placed sculpture in Roy Strong's garden.

Above: The black tracery of the oak tree's branches forms a distinctive silhouette against the winter sky. On mild days, the intricate pattern of its bare branches is an impressive sight in itself, but, under cover of frost, the effect is breathtaking – the smaller branches appearing like clusters of blossom against the dark outlines of the older wood.

Right: In winter, the sun is lower in the sky and shadows are at their most dramatic in the afternoon. The fine silhouettes of these trees, reaching upwards to the sky, cast long, dark shadows eastward at this time of day. Their blackness appears all the more striking because of the absence of distracting leaf and flower colour.

Your own garden will impose its own discipline, depending on its size, its levels and its setting. These, combined with your personal preferences, will determine the basic theme of your planning. Just as in architecture, where the size and number of windows bear an important relationship to the area of wall containing them, so the proportion between areas that are planted and those that are not, must be kept in balance.

Gardens that look and feel good in winter do not happen by chance; they are the culmination of designs first sketched on paper and then considered and perfected. I think they are best when they evolve slowly, because much of the happiness of owning a garden lies in its development – watching it grow into itself, thinking up new schemes, learning to observe space, structure and pattern.

Silhouettes and Shadows

Learning to garden is like most arts: it teaches you to be observant, to look at plants and remember them, so that their potential shapes are clear in your mind. To cultivate this memory is important – it means that you will be able to make a better choice of neighbours for your trees and shrubs or, if they are to be free-standing, to visualize their silhouettes and the shadows they create, and plan them accordingly.

Silhouettes of trees, when the sun is low in winter, stand so clearly defined against the sky that when I am driving through the countryside, I often challenge myself to recognize a tree by its outline. Elm, oak, ash and beech all have distinctive shapes, determined by the way in which their branches grow upwards, horizontally, or downwards from the trunk. The same is true of the trees and shrubs in our gardens.

The shapes of the trees you choose will dominate the background, command the middle distance, and provide the focal points for your overall garden plan. Bear in mind what the designer David Hicks recommends when deciding what kind of silhouette – pendulous, tiered horizontally, upright – will suit which position in the garden. Take a black-and-white photograph, lay tracing paper over it, and then draw your new planting thoughts on it; immediately you will see what is best and where. Use this method for the other key elements of your framework – the siting of clipped trees and important shrubs, the contouring of new beds, the routing of new paths.

For your garden you may prefer trees that are broad-

Right: Such a collection of conifers might appear uniform and uninteresting in summer but, through the haze of a frosty winter morning, the striking silhouettes, the variations in height and shape and the gentle gradations of greys and greens provide welcome diversity and colour. The snaking, mounded bed of heather and conifers creates an unusual and dramatic vista.

Below: It seems miraculous that phlomis seed heads on their fine brittle stalks are able to survive the worst of winter weather but, even after a fall of snow, they remain upright and undamaged, their warm brown colour dramatic against the snow-covered ground.

headed, like many of the acers. Sorbus range from broad- to round-headed; if it is round-headed crowns you want, then crataegus, *Gleditsia triacanthos* and most of the prunus and malus will be suitable. Some of the malus, however – *Malus* 'Golden Hornet', *M.* 'Red Jade' and *M.* 'John Downie' among them – are often so covered in fruit that their branches, though naturally round-headed, become pendulous. For a round-headed evergreen, I would choose the holm oak, *Quercus ilex*, grown on a 90 to 120 cm (3 to 4 ft) stem and then kept trimmed into shape. The evergreen Portugal laurel, *Prunus lusitanica,* can be given the same treatment.

The choice of weeping trees is sparser, but they all look good in winter because of their outline. *Betula pendula* 'Youngii' becomes dome-shaped, with branches eventually reaching the ground. *Cotoneaster* 'Hybridus Pendulus' has stiff, firm branches and a generous covering of red berries. The American weeping willow, *Salix purpurea* 'Pendula', is narrow in habit and has attractive, purplish, vertically hanging branches. At Barnsley we have the purple-leaved form of the weeping beech; it grows a thick cloak of branches and twigs, which even in winter makes a good hiding place for children. Under the branches of *Prunus* x *yedoensis* 'Perpendens' is a suitable place for planting spring bulbs. Neither this prunus nor Cheal's weeping cherry come into flower until spring, but both have distinctive winter shapes.

Columnar and pyramidal trees come in a wide range of species, both evergreen and deciduous. It is hard to choose evergreens without going to see them growing – either in a botanic garden or a garden open to the public where the trees are all named. Conifers vary so much in colour and density that you will have to study them and consider their silhouettes before deciding which will suit your garden best. The virtue of fastigiate deciduous trees is the small amount of ground space they occupy. My choices would be the Lombardy poplar for a tall, distant feature or for boundary planting, the balsam poplar for scent, and *Acer platanoides* 'Olmstead' for its erect green column in summer. A stand of them when leafless is impressive.

Among evergreen shrubs with attractive silhouettes, I would suggest *Prunus laurocerasus* 'Marbled White' and the low, compact *P.l.* 'Otto Luyken', the dome-shaped *Osmanthus decorus* (syn. *Phillyrea decora*) and the rounded *Hebe rakaiensis,* whose symmetrical outline will add form to the winter garden and which can be used in many places where a full stop or other emphasis is needed.

For deciduous shrubs with distinctive silhouettes and interesting skeletons, choose *Viburnum plicatum* 'Mariesii' and *Cornus controversa*, both with horizontally tiered stems. For stems that grow upright, giving a strong vertical line, use *Salix alba* 'Britzensis' (syn. *S.a.* 'Chermesina') and *S.a. vitellina*, pollarded close to the ground.

Some trees and shrubs are like some people, however – short on shape. Deciduous shrubs I can think of with less-than-model figures include deutzias, weigelas, kolkwitzias and *Viburnum x bodnantense. Chimonanthus praecox* has yellow flowers and is a fast grower, putting on 60 to 90 cm (2 to 3 ft) each year, and you will want to plant it close to the house to catch its scent as you pass by, but it is not one of the most elegant shrubs in summer. You can relieve its heavy appearance during August and September by giving it a clematis which requires hard pruning each year – I am always looking for new places to grow late-flowering clematis. Do not banish these shrubs from your garden – instead, place them where they can merge with their neighbours. Use the shapely ones as specimens and put them firmly in the forefront.

Winter shadows are more obvious than summer ones – long because the sun is low, they add a new dimension to the drama of winter gardening. On sunny days they look quite different at the beginning and the end of the day – changing in shape and direction, in intensity and sharpness of outline. Their density is affected by the material on to which they are cast – stone paving, brick, gravel or grass. The late Thomas D. Church, a famous American west-coast gardening writer, summed up the games that shadows play: 'Whether broadleaf or conifer, deciduous or evergreen, tree shadows are soft and clustered at noon, long and dramatic at sundown – needed for shade, welcome for pattern.'

I have an imaginative friend who is making a garden from a field of several acres around his new house in Philadelphia. He was planning, when I was there, to put up tall, free-standing archways at three points of the compass – east, south and west – to cast shadows towards the house and his flower beds. Three times a day the shape of the arches will appear as a shadow on the ground, outlining the shape of his flower beds.

Shadows can be more impressive than the objects from which they fall. At Gothic House in Charlbury, a high trellis-work fence and a single tree silhouette form an unusual combination, but it is the shadows they cast that reveal the full intricacy of the trellis design against the twisted form of the tree's branches.

Another friend, when she was planning a 'natural' pond, decided to site it near a large sycamore tree. One evening she was considering what size the pond should be when she realized that the sycamore's shadow was falling in exactly the right place and outlining the perfect shape.

John Ruskin had no doubt of the importance of shadows. In 1854 he wrote: 'They are in reality, when the sun is shining, the most conspicuous thing in the landscape, next to the highest lights.' Ruskin's observation was brought forcibly to my mind when I was walking along the yew tunnel at Melbourne Hall in Derbyshire. The surprise as I emerged from the dim light was to be greeted by warmth and sunshine. I remember the same sensation as I walked out of a 'tree house' in Lady Westmorland's garden at Rodmarton in Gloucestershire.

On clear winter days, shadows are crisper, and they really come into their own when the ground is snow-covered. Our lime walk runs from north-east to south-west; on sunny mornings, when I go out after breakfast and look across from the temple to the fountain, the shadows are lying flat across the grass, making a stage set of their own. By noon the sun has risen above the trees, and the magic has gone. The yew trees that march beside the path, at right angles to the lime walk, have shadows falling first one way and, by tea time, the other.

Look at the pattern of shadows cast in your garden. You will see them everywhere, some accidental, others contrived. A pergola, bare of the leaves of its climbers, naturally casts shadows in a regular series of hoops. A pierced wooden or iron gate throws an intricately patterned shadow onto the ground. At Stancombe Park, the shadow of the Chinese Chippendale handrail on the floor of the bridge, creates a memorable double image. Perhaps, like me, you agree with C.S. Lewis – 'This life is only shadows, real life has not begun yet.'

Allées and Vistas

Allées and vistas are the visionary elements in the design of any garden, small or large. They lock the garden into its landscape and control its views of that landscape. Allées are the thongs of formality – they bind a garden into its constituent parts. To a small garden they are as corridors are to a house; to a large garden they are the ley-lines of its design. Vistas are the hidden breadths of vision – the views that no-one suspected the garden had. They open out, now here, now

Above: In Roy Strong's garden, the dramatic Elizabeth Tudor walk is a double allée, a grass walk flanked by pleached limes and Irish yews. In spring thousands of pale daffodils stop you in your tracks but in winter you feel compelled to walk into the distance towards the Shakespeare urn placed against a backdrop of darkest evergreens.

Right: At Wildenborch in the Netherlands, the structure of the garden is revealed by winter's hand. Proportions are carefully and skilfully balanced. Against a background of ancient yews, a high beech hedge has been tightly clipped to give the impression of an inner wall and trained into sculpted arches to link with an outer hedge of evergreen topiary. Winter shadows fall gently on the statues which mark the entrance to an axial pathway.

there, to reveal the landscape outside and beyond the garden or other parts of the garden itself.

Vistas and allées, and the difference between them, were forced on my attention by the late Lord Buchan; when walking round my garden with me, he said casually, 'I like the way you make vistas into your borders.' I had to think what he meant, then realized he was referring to my hopeless way of allowing tall plants to come in front of low ones – something we probably all do until we learn more about the habits and heights of herbaceous plants and shrubs. Traditionally, borders have low plants at the front and are graded higher and higher towards the back. My planting was not very well thought out, so you sometimes got views through to the back of the border.

This chance remark led me to think for the first time about vistas and allées. What is the difference between the two, I wondered? An allée is formal and an integral part of a formal design. It is a pathway, usually flanked by symmetrically planted deciduous or evergreen trees, sometimes by clipped hedges as well. It has an entrance and an exit directing you from one area to another, and it often leads you from one axis to another. It may terminate with an eye-stopper, send you off in another direction, or simply invite you to walk on into the distance. To finish an allée with a wall and a climbing rose would be wanting in imagination, an opportunity wasted.

A vista is essentially a view (*vista* is the Italian for view). It can be contained formally – in fact along an allée – or it can be created informally. Whereas an allée is always straight and

relatively narrow, a vista can be broad or narrow. It can have a single focal point or open into a whole landscape. There can be a direct route to it or it can be cut through at random. It can be within the garden or can incorporate the landscape outside.

Allées are actively enhanced by winter, whereas vistas tend to lose some of their impact. In summer the strict formality of an allée inevitably is softened by leaves, and its grass floor is just one green among many. In winter the soft ribbon of lawn acts as a foil for its brown deciduous walls, but if its hedges are evergreen, the architectural impact of the allée is even more forceful. Vistas, on the other hand, suffer, because they have forfeited their element of surprise – when every tree and hedge becomes see-through, the eye is no longer caught. The landscape everywhere is open to view.

Walk along the hornbeam allée at Hidcote, and as you arrive at the far end, you will be arrested by a breathtaking vista – the view that persuaded Lawrence Johnston to buy the property.

In my own garden, the lime and laburnum walk, the principal formal feature in the garden, is planted as an allée, ending with a stone pillar. Halfway along the allée and at right angles to it is an incidental vista, looking through a border and an opening in the yew hedge to a stone statue in the distance.

Allées and vistas are for strolling in in summer, for looking at and enjoying from a distance in winter. I get as much pleasure in winter from looking out of my window at the sea of red twigs on top of my lime allée as I do from walking down it at the height of summer.

Vistas can be accidental or planned. They can suddenly be opened up by cutting down a tree or cutting 'windows' through a hedge – clipping leaves and small branches to form an aperture giving a new and often unexpected view. At the Ladew Gardens in Maryland, openings in the thuja hedges give dramatic views of the countryside and beyond over distant hills. 'Windows' can be just as effective in a smaller garden where you want to have a glimpse into the next 'room'.

Round windows and doors are far more likely to remain in one's memory than the conventional rectangular kind. In Mrs David Rockefeller's garden in Maine, the 'moon' gate caught my eye from a distance and the nearer I got, the more I saw the evergreen tree beyond as a focal point. In winter, with little planting in the foreground to distract the eye, the gate is even more effective. At West Green House near Hartley Wintney in Hampshire, there is a round gate in the red brick wall leading from a newly planted box garden into the rose garden. It is an enticing view. And Lady Anne Cowdray, in her garden near Devizes in Wiltshire, has a circular *clairvoyée* in her brick wall, through which the sunken rose garden can be glimpsed.

You can create a vista by lining up a series of arches to a focal point – as at Cambo in Northumberland – or simply by making a path leading into the distance. I have never been to Powis Castle in the winter, but the vistas along its terraces must look exciting with the straight beds edged with clipped box, 30cm (12in) high. At Barnsley, the gate from the greenhouse yard looks directly across the garden to the Gothick summerhouse; a yew hedge on one side and a rose hedge on the other lead your eye to the building, especially in winter when the nearer borders are leafless and low.

A vista need not be straight, however. It can have a curve to entice you to walk on and discover what lies beyond. Curves are most commonly used in tunnels and pergolas. Candida Lycett-Green, daughter of the late Sir John Betjeman and herself an author, created a gentle sweep in a potager she made, with roses and clematis clambering up arches. The laburnum tunnel at Airlie Castle in Aberdeenshire winds slowly to the left. When I first read about it, I wondered if I would like it. In reality, it looks wonderful. It was not designed with winter in mind, but the repetitive rhythm of the tracery of bare branches is interesting, and late winter bulbs could be planted to give colour in February. One of the best known and most formal of curved pergolas is alongside the Pinmill at Bodnant – a gentle curve of architectural trelliswork with the added drama of a change of level at its end.

An allée must end in a focal point – a summerhouse, church tower or temple, an obelisk or pillar, an urn or statue set into a niche or a curved hedge – or it can bisect, acting as the pivotal point for another allée. Whichever type it is, it should stand out against its background, be it clear sky or dark hedge. St Paul's Walden Bury in Hertfordshire has no less than five allées, three of them with a cross-axis, each ending in a view – of the parish church, of a statue, of a temple, across the lake, or, looking back towards the house, of one Adam wing.

Remember when you create an allée that you can use the effect of perspective to make things appear taller or shorter. At a distance of 27m (90ft), a 1.2m (4ft) obelisk on rising ground will appear to be at least 1.8m (6ft) high; conversely, when you plan something at the lower end of an allée, it will tend to look smaller and squatter. When Simon Verity carved a hunting lady for us and we sited her to be seen through a break in the yew hedge, for the first time I became aware of a rise and fall in the lawn – to see her feet she had to be placed on plinth.

There are also ways of increasing the length of an allée visually. The trees or objects you use can be planted to be smaller as they recede: a normally man-sized object such as an archway need only be 1m (3ft 4in) high, as long as you do not expect to go through it too often. The open-air theatre at the Villa Marlia near Lucca in Italy is a serious example of this art of deception: dark green, closely clipped wings focus

The winding pergola in Lord Aberconway's garden at Bodnant in Gwynedd, north Wales, is both a vista and an allée, with views out to the sides and along the length of the tunnel. The low, curved trellis arches and longitudinal beams which support summer growth, in winter cast shadows on the ground and are widely spaced enough to allow dappled sunlight to penetrate.

At Parnham in Dorset, home of master furniture-maker John Makepeace and his wife Jennie, allée and vista are skilfully combined. Punctuating the smooth expanse of frosted lawn in front of the house are parallel rows of regularly spaced, clipped pyramids of yew, leading away from a double flight of steps into the park to form an axial ride. Although a simple design, the effect has both drama and elegance, reminiscent of the terraced foreground of a French Renaissance château.

the eye onto the backcloth of the raised grass stage. The backcloth is a tall yew hedge pierced with three arches, in each of which sits a statue. It is not until you climb onto the stage that you realize the figures are diminutive and that you will have to stoop to pass through their arches. The enclosed, totally green circle of stage and auditorium wings stand at the end of a long grass allée.

Colour, too, can heighten the impression of length. Bright colours such as white and yellow stand out and look closer than dark colours at the same distance, so leaf colour in the distance should be at the quiet end of the spectrum – deep purples, heather colours and blues.

Herbaceous borders are sometimes more dramatic if they are planted as 'twin' borders. They then constitute allées. At Bramdean House in Hampshire you see them through an iron gateway; the grass path between them leads to the house. At Rodmarton Manor in Gloucestershire, at the end of the allée created by the twin borders, the focal point is a steep-roofed summerhouse. The borders are punctuated halfway along by a round pond and yew trees. I love this view in winter when the colour of the stone summerhouse stands out and the pond invites you to sit on its raised walls. The well-known borders at Bampton Manor in Oxfordshire are dramatic in winter, backed with yew hedges and higher pillar-yews at intervals and focused on the church steeple in the distance. Even the most modest of cottage gardens can have an allée from its gate to its front door with twin beds each side of the pathway.

Allées on the grand scale in parkland, like those at Badminton and Hampton Court, will always be exciting. Less well known perhaps is Harold Peto's modern interpretation of the dramatic watercourse at Sceaux in France. At Buscot Park in Oxfordshire he has created an allée with a narrow watercourse along its entire length, flowing in a series of steps towards a distant lake. Flanked, first by grass paths, then by low clipped hedges and finally by bosquets of trees, it is an essay wholly in the spirit of 17th-century French gardens.

George London and Henry Wise, the leading garden designers of the early 18th century, laid out a wonderful garden at Melbourne Hall in Derbyshire with vistas stretching over lawn and water towards woods. The allées radiating out through the mature trees must be more impressive today than when the garden was first laid out.

You may be sceptical that ideas on such an ambitious scale can have any relevance to our small gardens of today. There are, however, ways of creating allées and vistas even in the most modest garden which can be enjoyed without too much extra work. The late Robin Spencer's garden at York Gate near Leeds is a model of intricate design within a small space.

As well as using twin herbaceous beds to make an allée scaled down to cottage-garden size, you can also create small allées in an intimate area of the garden by lining a path with low shrubs neatly trimmed. Box balls or pyramids are ideal; so are dwarf conifers with their tidy habit of growth. We have box balls in the so-called spring walk; in winter when there is

very little greenery about they come into their own. In summer the balls are almost concealed by the various herbaceous plants. If, on the other hand, these clipped shrubs, set in the lawn, are used as punctuation marks along a drive, they remain in evidence all through the year, not merely as something solely for winter. If you want to achieve this effect, it is important to use a shrub which you know to be completely hardy in your area; a formal statement that relies on the rhythm of repetition can be spoilt by the death of one along the line.

Architectural Details

The architectural details of the garden – paths, walls, steps, terraces, statues and urns – play a greater role in winter than in summer. Their main function is to ease the transition between one area of the garden and another, particularly where sharp changes of level occur. They also serve to signpost exits and entrances, to mark changes of direction and provide the permanent containment of parts of the garden needing protection.

Architectural details ought never to dominate the garden, but always to enhance the textures and features of nature. In winter the walls that supported the clematis, the stonework that was hidden by the voluptuous summer growth, the symmetry of the garden design, emerge as structural features in their own right.

Paths may range from a simple ribbon, to a well worn track leading through a heather garden, a grass swathe mown through a woodland area, or a geometrical brick and flint design. When the general greenery of a garden has died down, the patterns of its paths stand out boldly. The colour of the materials should blend with local materials, but even in stone country, brick can sometimes be effective. Grass paths look well with an edging of brick or cut stone. Hard paths can be checkered – alternating stone with squares of gravel, brick with squares of flint. You can devise endless patterns, and they are such an enjoyable part of life. People who do not appreciate patterns really miss something. It is the impress of pattern that will hold the design of your garden together – especially in winter, when you are more conscious of it.

Because patios and paved areas are permanent features, it is important to spend time and thought on the material you use. Look in books and other gardens for ideas. Choose a

paving texture that will be easy to walk on, and a pattern that will suit the house. Given free choice, and an unrestricted budget, I would always use York stone in preference to any man-made product, especially if it can be cut into the shapes you want. It looks wonderful, but can be slippery when wet. Stone can be combined with bricks laid as an edging to each slab or used in a basket weave pattern. Crazy paving is one of my dislikes and has nothing to offer for winter design.

In summer, walls are usually covered with plants, and terraces and tubs billow with them, but in winter there they are for all to see. You may be lucky and have an ashlar wall like my 18th-century one. Or it may be rubble, or brick. If it is to be a new one, use local materials whenever you can. Walls are expensive, but once built need little maintenance. Make sure that they are properly buttressed or have expansion joints at regular intervals. Think about the need for pillars, which add strength and decoration, and about the need for sweeps to lower sections. Most important, remember that your wall needs to be capped properly to protect it from frost and rain. All old walls will be a constant joy and a marvellous place for climbers. For thoughts on winter climbing plants with which to clothe them, turn to page 53.

What will you do about steps? They can become more than just a means of access – they can be decorative and will be at their most noticeable in winter. They must be shallow enough to walk up and down with ease. The risers, I believe, should not be more than 11cm (4½ in) high – (the official recommendation is 11 to 15 cm (4½ to 6 in) – and the tread of each step at least 30 cm (12in) deep. A narrow stairway looks uncomfortable and difficult to negotiate, so make the steps as wide as you can, and if they lead down to a lawn or a terrace give them an elegant line, increasing in width as they descend. Circular steps with a landing halfway up are generous in feeling, and an elegant solution.

If your steps have a retaining wall each side, this is a splendid place to plant shrubs which will tumble, and others, such as lavender and rosemary, whose scented leaves reach upwards to your touch. Plants soften the contours of a stairway, and to maintain this effect in winter, use a certain number of evergreens – ivies, hebes, vincas, *Lonicera pileata,* with *Iris histrioides* and *I. reticulata* planted round them. If the steps are at the end of a low straight wall, change the mood by giving them a curve. This could be the place to plant your *Helleborus orientalis,* so that you can look into their hanging faces as you go up the steps.

Left: Frost and silvery winter light transform the Pinmill terrace garden at Bodnant into a stage set. All trace of rich summer planting has vanished to reveal the fine detail and disciplined simplicity of the garden's design. Wooden benches, trelliswork fences, old stone buttresses and evergreen pyramid shapes add incident to the flat expanse of terraced garden.

Below: Permanent features give structure and incident to the garden in winter. In my own garden, an old stone wall covered with the greens and golds of Hedera colchica *'Sulphur Heart' provides the perfect setting for Simon Verity's statue. A sprinkling of snow and afternoon sunlight emphasize the silhouette, the curves and the shadow of this gardening lady with her basket of flowers.*

Right: The threads of a knot garden are best kept simple. At Barnsley I have used an interlacing pattern formed of box and wall germander, infilled with different gravels. The corners are marked by slightly taller, rounded box balls. The four Ilex x altaclerensis *'Golden King' beyond the corners provide accents of height which link the knot to the house and garden.*

Below: In winter months, the structural features of the late Robin Spencer's garden at York Gate, near Leeds in Yorkshire, are revealed. The gravel path, clipped hedges and geometrical shapes become more eye-catching than when softened or obscured by an exuberance of lush foliage and flowers.

Sloping lawns in a formal garden are generally uncomfortable. It is much better, if you can, to terrace the ground. Provide plenty of detail and interest for a terrace you walk along fairly often using evergreens and winter flowers. If your terrace is principally something to be viewed from above, it will be best to reduce your planting to two or three varieties, in groups large enough to make an impact. The top of the terrace wall will naturally be well drained, so choose low shrubs which like those conditions – lavender, dianthus, cistus and other 'greys'.

Statues and urns serve an architectural role in garden design, so their proper siting is vital. Placed well, they can form the focal point of an allée, or the central feature of an enclosed garden. They can be free-standing or backed closely by a yew hedge to set them off, or they can repeat themselves along the top of a wall. A pair of urns or statues can mark the way through a gate. Urns are happiest on a plinth or a wall; statues tend to dictate their own height. When you are siting them, think about the direction of the sun, and take into account the winter months when the sun is lower in the sky.

Your garden statues will be a matter of personal choice: like the pictures in your house, they may have been in your family for generations, or you may have chosen them yourself. I am very attached to the hunting lady and the dwarf gardeners carved for us by Simon Verity. I want to be able to see them often and easily. Garden statuary can be expensive; fortunately, there are many excellent reproduction urns and statues available today, which weather remarkably quickly.

Knot and Herb Gardens

Knot gardens, with their neat disciplined three-dimensional strapwork, are to me the perfect embellishment for any garden in winter or summer. Sir Francis Bacon, writing about gardens in the early 1600s, did not think much of them: 'As for the making of knots, or figure, with divers-coloured earths, that they may lie under the windows of the house on that side on which the garden stands, they be but toys: you may see as good sights many times in tarts.'

Gardens set out in geometric patterns made with the continuous threads of low, clipped evergreen shrubs or herbs became fashionable in the 16th century. With the country at peace, merchants and country squires filled their houses with

handsomely carved oak furniture and wainscotted walls, with panel paintings and decorative plasterwork ceilings. As the ladies embroidered – you need only study the material of the dresses in their portraits to see how much of their time was spent with their needles – they must have looked outside and wondered how pattern could be used to make their gardens more interesting.

Books of the period give a series of elaborate patterns for knots. *L'Agriculture et Maison Rustique*, translated into English by Gervase Markham in 1616, shows how a knot pattern, whether or not it was interlaced, must consist of continuous threads. He describes non-interlacing as 'broken quarter', while John Parkinson in 1629 used the term 'broken knots'. The 16th- and early 17th-century books suggest using lavender, germander, thyme, rosemary and other low-growing herbs to make the threads, emphasizing the interlacing of the overs and unders by contrasting colours. With far fewer varieties of flowers than we have now, the design of their planting was important. Both Markham and Parkinson agreed on infilling with herbs – low growing herbs such as pennyroyal, rue, chamomile, basil – and with violets and daisies. A taller herb or clipped tree emphasized the centre. Parkinson suggested using 'outlandish flowers' – those which had recently been introduced from abroad – to decorate the knot in spring and summer.

When threads were interlacing, the ground between was infilled with coloured earth. The colours suggested were all those used in heraldry, which leads me to think that the earliest known patterns were based on heraldic designs. The five colours were red (broken bricks, finely crushed), yellow (clay, sand or Flanders tiles), black (pure coal dust), white (chalk) and blue (coal and chalk mixed).

Today, using the same principle, different coloured gravels, which are cleaner and easier to use, set off the pattern very clearly. In southern Europe and countries where the light is more intense, especially in winter time, pale coloured marble chips are an alternative.

My herb garden is sited just outside the kitchen door. Designed as an elongated series of diamonds beside a narrow path, it makes a bold statement instead of blending unobtrusively into a border. Low clipped boxwood defines the compartments, which house a mixture of perennial and shrubby herbs, chosen to provide flavours for the kitchen throughout the year: parsley, sage, thyme, rue, hyssop, sorrel, golden marjoram.

As well as in 17th-century books with specific knot patterns, you can find ideas for knots all around you – on old book bindings, Jacobean plasterwork ceilings, Anglo-Saxon manuscripts, Celtic brooches, oriental carpets and the embroidered clothing of Tudor portraits.

The first essential when you are making a knot garden is to give it a suitable site, preferably near the house. It must be flat, and to get the full benefit of the pattern you should be able to look down on it from an upstairs window, or from a higher level. The raised walk round the reconstructed knot garden at Hatfield House in Hertfordshire is a perfect example: first you stand above the knot, enjoying its design, and then you step down into it and savour the planting. The pattern is clipped box, and the infilling solely plants and flowers which would have been available in the early 17th century, when John Tradescant was head gardener to Robert Cecil, first Earl of Salisbury. You may not have the ideal setting of Hatfield or the even older Bishop's Palace as an imposing background, but you can design a simple interlacing knot in such a position that you will see it from your house, winter and summer.

For best winter effect, I recommend a simple pattern with interlacing threads of box and wall germander, infilled with gravel or different textures or colours for contrast, and with a higher, clipped evergreen bush as a central feature or to accentuate the corners.

Make the pattern as simple as possible; do not be too ambitious, for the whole beauty of a knot lies in the perfection of its tending. To keep it in good trim is a lot of work. My own knot garden consists of two patterns, both with interlacing threads. Originally we used herbs, but soon discovered that they grew unevenly, and were impossible to keep looking neat and trim, so we discarded the lavender and feverfew and thymes, and used box of different shades of green throughout, except for one thread of wall germander, *Teucrium chamaedrys*, which clips satisfactorily, and a central feature of phillyrea. The effect is quite pleasing; as the box grows taller, the gravel becomes less visible, and the whole pattern is given movement by the rhythm of the interlacing.

To me, herbs mean beds and herb beds mean patterns; outlined in clipped box, they remain trim and interesting throughout the winter, while the herbs themselves are mostly below ground or not looking their best. There is no substitute for box when it comes to edging: wall germander, santolina, lavender and southernwood can all be used, but invariably need to be renewed after a few years, while box is ageless.

Before you start to design your planting, it is a good idea to sit down and make a list of herbs you like most, and allot them their space accordingly. Note their heights and whether they are evergreen (bay, sage and rosemary), and allow space for the summer-flowering annual herbs (marigolds and nasturtiums, dill and chervil). Remember to include enough paths or stepping stones, so that you do not trample the soil.

Obviously the pattern of your herb garden will depend on what space you have available. I have chosen to make mine into a series of tangent diamonds, each one edged with box. A different idea is to have checkerboard paving with herbs planted in alternate squares – a medieval pattern. Whatever design you choose, you can emphasize the corners with box balls or pyramids, upright junipers or rosemary.

You need not confine yourself to straight lines. I know a very pretty semi-circular herb garden radiating from the kitchen door. It looks like half a wheel, with stepping stones for its spokes. Herbs billow out over them in summer; in winter your eye is taken with the paths and shrubby herbs.

Not everyone wants their herbs in a formal setting: you may instead wish to incorporate them, cottage-garden style, into your beds and borders, mixing sweet-smelling honeysuckles and roses with culinary herbs. Purple and grey sage, golden thyme and feverfew all look at home in a mixed border, winter and summer. You could add an element of formality by making a seat with a trelliswork arbour and bay bushes as sentinels in the centre of two borders, edged with lavender and rosemary for winter.

A formal herb garden will fit best near the house, while a more relaxed arrangement should be sited further away – but not too far or the hurried cook will not have time to fetch the parsley and mint.

For a winter backcloth, the herb garden could be outlined or backed by a low hedge of shrubs with attractive foliage, kept well clipped. Chaenomeles might be grown with *Pyracantha atalantioides* (syn. *P. gibbsii*) in summer entwined with honeysuckle; *Hebe* 'Autumn Glory', which is evergreen and flowers through until December, could be combined with *Skimmia japonica* 'Rubella'. A low rosemary hedge is another possibility and always a joy: think of the pleasure of laying your pillow cases and handkerchiefs, even your tea towels, on it to dry and soak up the scent. But be warned: the hardiness of rosemary and many of the Mediterranean-type sub-shrubs depends as much on soil conditions as air temperature. Damp will be fatal; consider their natural habitat.

Skimmia japonica 'Veitchii', a female form of the excellent shrub that provides an evergreen backcloth to the herb garden, has large panicles of red buds in winter, later opening to white, yellow-anthered flowers in early spring.

Topiary

Topiary, one of the most ancient of gardening arts, looks as decorative in a winter landscape as a knot garden does. Evergreen may be clipped to make an impact, or elaborately sculpted to create a fantastic shape.

You may limit yourself to a pair of box balls, spirals or pyramids each side of the front door, which you will tend little and often with your nail scissors, keeping them as perfect as you can through all the year; in winter they will be your pride and joy. You may extend to an occasional clipped bush to

Human footprints and animal tracks in the snow reveal that the privacy and seclusion of this intimate garden have been disturbed. A fine geometric layout and bold topiary shapes are the basis of this garden in the Netherlands, designed by Jaap Niewenhuis and Paula Thies. The garden's formal structure is emphasized and given luminosity by the light covering of snow.

accentuate a special point in the border — the corners of a garden room or your rose or herb beds. Or you may be much more ambitious, devising a series of intricate shapes for one area of your garden or even a whole garden room. The possibilities are endless.

You may even accept William Lawson's advice in *A New Orchard and Garden* in 1618: 'View now with delight the works of your own hands ... your trees standing in comely order which way soever you look ... Your Gardener can frame your lesser wood to the shape of men armed in the field, ready to give battle, of swift-running Grey-hounds, or of well-scented and true-running Hounds to chase the Deer, or hunt the Hare. This kind of hunting shall not waste your corn, nor much your coin.'

From the 1930s until World War II, Harvey Ladew, a keen hunting man from Baltimore, Maryland, travelled to England each winter to hunt, and created back home a topiary hunting scene with huntsman and hounds galloping across the field in full view of his house. At Knightshayes Court in Devon a topiary fox runs along the top of a yew hedge with hounds in full cry. Shapes such as these may take years to reach perfection, but it is surprising how, with ingenuity, they will soon become recognizable as the fantasy you had in mind.

If you are to create topiary, you have to think which trees and shrubs are best suited to clipping and shaping. Obviously those with small leaves can be made into the neatest shapes. They may be evergreen or deciduous, but for winter effect you should choose those which keep their leaves. Yew and box improve in quality with age, whereas the faster growers will be more difficult to control. You can find *Taxus baccata*, the common English yew — the best variety for topiary — at most good garden centres.

Study the shapes of your plants carefully in order to decide what you can ultimately make of them, and allow them to settle before starting to clip. I love the yews in the 'pillar garden' at Hidcote, but so ambitious a scheme must be a long-term project. Did Lawrence Johnston copy this idea of 'cubes' of

Ladew Gardens in Maryland are world-famous for the topiary hunting scene, complete with horse, huntsman and hounds. The brilliantly sculpted forms, clipped out of Japanese yew, make a statement of absolute bravura, standing out against the leafless trees in winter like outsize chessmen. Red ribbons add a touch of colour at Christmas that looks striking against the dark green of the yew and the snow-covered ground.

yew, each one surmounted by a pillar, from the 1745 engraving by Highmore and Tinney of the Hampton Court garden?

Nathaniel Lloyd, in his book *Garden Craftsmanship in Yew and Box* (1925), by far the best on the subject, demonstrated how to treat well-grown yew trees and turn them into topiary. If you are moving into an old garden and by good fortune have an overgrown yew hedge or one you want to reduce in size, think carefully before working on it – you may be able to save taller shoots at intervals and turn them into pyramids, peacocks or balls to surmount the top of the hedge.

Box is perhaps not as versatile as yew to mould into shapes; conventional forms such as cones, pyramids, balls and spirals are the most appropriate. Their simplicity adds style to a formal planting. When choosing your specimens, you must decide on the height you are aiming for and use either *Buxus sempervirens* for larger results – up to 3 m (10 ft) high – or the dwarf *B.s.* 'Suffruticosa' if you want quite small examples.

I am an avid collector of box, of which there are many different cultivars and species. When Lawrence Johnston died, he left his famous garden in the south of France to Nancy Lindsay, who brought cuttings of many of his plants back to England. The plant I treasure most from Nancy is the striking golden-leafed *Buxus sempervirens* 'Latifolia Maculata' (syn. *B.s.* 'Aurea'). The pattern of pale green and yellow is different on each leaf, but the overall effect throughout the summer is of gold, turning to an exciting bronze which lasts through the winter. Where we have used it as an edging in our vegetable garden and trimmed it hard, it is a glorious golden colour, especially on the young growth.

Another favourite is the cultivar *Buxus sempervirens* 'Elegantissima'. Many years ago I bought two pots of this at Mount Vernon, George Washington's home on the Potomac River in Virginia. I like the association and wonder whether they may be descended from an original plant he had there. We have made these and several cuttings from them into balls to stand each side of the gate leading into the vegetable garden.

Another, more recent, moment of excitement for me was to see a line of fastigiate box, quite 1.8 m (6 ft) tall yet little more than 30 cm (12 in) in diameter, growing in a garden in Charlottesville, Virginia. I was assured that they were neither trimmed nor tied to keep them in this perfect columnar shape. The garden owner kindly gave me some cuttings; some others she had given to the National Arboretum in Washington DC were 90 cm (36 in) tall when I saw them, and I was given more cuttings from there also. Now my propagating frame has thirty cuttings, carefully guarded and all rooting. They could well become an attractive substitute for some of the upright conifers which so many people plant in England.

I prefer the Irish juniper, *Juniperus communis* 'Hibernica', to the various columnar cypresses. Their habit of growth is denser and to me makes a more effective weight – my only reservation is that junipers often begin to look old and ragged after about twenty years, so one should keep young specimens growing on to replace them when necessary. They are not strictly speaking topiary, since apart from a very occasional tidying of wayward shoots and some cutting to prevent them growing beyond the desired height, they need no attention.

I have used the Lawson cypress, *Chamaecyparis lawsoniana* 'Ellwoodii' in a line, alternating with mounds of grey santolina, in front of a terrace wall on our driveway. Again this is not topiary, but the trees all need tying round to prevent snow getting into them and an annual clipping to level them off, if, as I do, you want them to have a flat top.

You may well find a shrub already established in your garden which you can transform into a ball or a pyramid, or into tiers. In medieval times, artificial contraptions called estrades were made from flat, circular metal baskets filled with flowers attached to a vertical pole – an idea probably brought to Europe from Buddhist countries, where stupas or relic mounds would be surmounted by triple 'umbrellas'. In Buddhism an umbrella symbolizes royalty – the king alone was allowed to sit in its shade. The three tiers – in Sanskrit the word is triratna – represent the three jewels of Buddhism, their three vows: 'To the buddha for refuge I go, To the samgha for refuge I go [the monastic community], to the dharma for refuge I go [the teaching].' As you clip your tree into three tiers, remember it symbolizes the tree of enlightenment beneath which Buddha sat; in your turn, sit beneath it and contemplate the truth about your garden.

'Hedges on stilts' are historically a long-established and most effective form of topiary. My red-twigged lime walk (*Tilia platyphyllos* 'Rubra'), made many years ago, is now well established. The limes are cut hard into shape – short back and

A 'hedge on stilts' is a jeu d'esprit *among topiary shapes. It forms a light, elegant screen, casting interesting shadows, giving dappled shade and allowing views between its trunks. At Barnsley, the crimson haze on the avenue of limes,* Tilia platyphyllos *'Rubra', created by the young growth on the leafless red twigs is most dramatic when lit up by the early morning or late noon winter sun.*

sides – in early June, and the new growth tidied again in August or September, when the twigs on top are left to form an amazing red haze which is quite one of my favourite winter sights. Each spring I have to make the firm decision to cut them hard back again as soon as the sap is rising, to start the whole cycle once more.

In another Gloucestershire garden, around a converted barn, limes have been planted in a quincunx, three deep, to act as a protection from the road. It is hard work keeping them under control, but the pleasure is year-long.

The pattern of the tracery of the leafless twigs on pleached hornbeams can also be wonderful, especially when a light covering of snow shows them off to advantage. You can see these at Hidcote, where the double banks of trees are planted as pairs, and also at the late John Fowler's home at Odiham in Hampshire, where they are planted singly in a line. At Dumbarton Oaks, Washington DC, Beatrix Farrand planted a double file of American hornbeams in a circle, with a path leading you around the two rows of trees. A different treatment is the series of hornbeam arches growing against the terrace in front of the Château de Limpeville in France; in winter they look like oversized leafless croquet hoops.

Any tree which responds to clipping and pruning can be used to make hedges on stilts. At a small country house in my village I have enclosed a tiny formal garden outside a drawing-room window with *Sorbus aria* 'Lutescens'. Now three years old, it is starting to give the required effect. In summer the soft grey leaves form a light screen on stilts; in winter it is the pattern of the interlacing branches which is attractive.

The most useful of the taller evergreen topiary trees is *Prunus lusitanica*, the Portugal laurel, which became fashionable again in 1912 when it was shown as carefully clipped 'mushrooms' on 1.2 to 1.5 m (4 to 5 ft) stems at the Chelsea Flower Show, by a firm appropriately called Cutbush. To me it is the epitome of elegance: it sparkles with its bright green leaves in winter and may be clipped hard to make perfect domes or any shape you will. It could just as easily be turned into a hedge on stilts.

The variety *P.l.* 'Myrtifolia' has a pyramidal growing habit and can be a hardy substitute for a bay tree in a tub. We have one plant of *P.l.* 'Variegata' here at Barnsley, growing as an ordinary bush; it always draws comment from connoisseurs, who know how difficult it is to find. Imagine what gasps it would bring if you had a carefully clipped line of this variety of the Portugal laurel growing in Versailles tubs in front of your house. A daydream perhaps, but one which nurserymen should not rule out. We watch our children grow, so why should we not our shrubs? The true gardener has patience; we must not forever be craving an instant garden.

Holly grown on a stem has been part of the topiarist's stock-in-trade for generations. Today plenty of nurseries sell good specimens which you can then torture into shape with your secateurs. It is a shrub you must look at carefully before deciding how you are going to shape it. If it is pyramidal, with branches growing horizontally, then it will be ideal for clipping into tiers. If it is round and copious, then it will make lovely, voluptuous single balls or series of balls climbing up the central stem. You must just stand back, observe and allow your imagination a free rein. Then you will come up with the right answer for that particular tree.

Quercus ilex, the evergreen oak, is used as a conventional tall hedge more often in Italy than in England, but considering it has survived for centuries in this country we should be more aware of its possibilities. Visit Hatfield House in Hertfordshire and see these oaks on 1.8 m (6 ft) stems, clipped into square heads, lining the west garden. For a more modest garden, a pair would look handsome flanking a gateway or as features each side of the exit from one garden room into the next. I can also imagine them as a neatly trimmed evergreen archway.

On a smaller scale, two shrubs for quick-result topiary are *Ligustrum ovalifolium* and *Lonicera nitida*, in both their green and golden-leaved forms. They grow quickly but are apt to be floppy if they are not attended to; clipped hard, especially the lonicera, they make good shapes in a very few years. We have golden privet as standard balls (a suggestion by one of my best friends, the gardening author Peter Coats) marking the corners of some of the beds in our kitchen garden and now 90 cm (36 in) higher and two years later, they do their job well. Our small standard golden loniceras are in pots, which restrict their roots and slow down their growing pace. If you leave the clipping until spring, just before growth starts, you will have amusing golden heads, looking rather like punk hair-dos all winter long.

Another shrub you could try as a standard clipped into a ball is *Euonymus fortunei*. There seems recently to have been a supply of these coming into garden centres. They are grafted onto 90 cm (36 in) stems and it is important to make sure that the graft is in good condition. The heads will take two years to thicken, and it is up to you to keep clipping the new growth to make a good solid globe. You can use them in containers or tubs, or grow them free-standing on the edge of your lawn or flower border.

These ideas are all geared to English gardens, but gardeners who live in colder or hotter climates can choose trees and shrubs more suited to their own environment – rosemary in the south of Europe, the naturally pyramidal *Ilex opaca* 'Arden', hardy in the USA in all but the most frozen states, and exotic hibiscus in tropical climates.

On the east coast of the USA there is a fashion – more indoors than in the garden – for making topiary shapes from small-leaved ivy and ficus trained over wire frames and filled with moss to retain moisture. It looks quite easy to do, but in fact it requires considerable expertise. To make really good specimens is an art.

Another evergreen thought, illustrated in 19th-century books by Jane Loudon and later by Shirley Hibberd, is to cover archways and arbours with ivy for the winter. Kept well clipped, they give the appearance of topiary. I would grow summer-flowering clematis with and through the ivy, choosing those which need to be pruned back hard in spring – *Clematis* 'Jackmanii Superba', *C. tangutica*, or any of the Viticella clematis.

Tapestry Hedges

Hedges combining shrubs of different colours and textures consist mostly of evergreens and are therefore of special value in winter, both for protection and for brightness. Whereas the unity of a single-species hedge gives you pleasure when it is closely clipped and tidy, a tapestry hedge offers change of leaf shape, size, texture and colour. Differences in the rate of growth and the difficulty of even clipping are the only drawbacks – a yew or box hedge will look trim from an early age, but a tapestry hedge may be somewhat uneven and imprecise until its various constituents become united.

Which shrubs grow best alongside each other? Box and holly in its green and variegated forms will be the mainstays, together with yews, *Taxus baccata*. All are reliably hardy. Of the three, I think yew grows the fastest, but they all require hard clipping after their first year to maintain a good solid effect, with each shrub well-clothed to the base. The thick, tough, dark green leaves of *Buxus sempervirens* 'Handsworthensis' make this form of box excellent for tall hedges and screens.

At Barnsley, golden privet, variegated holly, and tawny beech line the grass pathway leading to the gothick summerhouse. Clipped to distinctive shapes to complement the shape and texture of their individual foliage, they create a memorable array of winter colour.

The beautiful sixty-year-old hedge at Rodmarton Manor in Gloucestershire – of box and yew with holly cones – is a joy throughout the year and its solid structure fills you with security. The total effect is exceptionally pleasing; with its smaller leaf, the box gives a touch of lightness. I like the simplicity of using just three shrubs.

The Rodmarton hedge has a splendid 'batter'. As Nathaniel Lloyd explained in *Garden Craftmanship in Yew and Box*, a hedge is meant to represent, or be a substitute for, a wall. If the sides slope out towards the base, creating a batter, the lower branches get sufficient light to keep their fullness; if not they may well become weak and straggly.

Other evergreens you can use for hedging are *Osmanthus* x *burkwoodii*, with its smaller pointed leaves, and *O. heterophyllus*, with its glossy, spiny leaves. Both are more unusual than the Portugal laurel. I am also keen on *Rhamnus alaternus* 'Argenteovariegata' for winter; the leaves, marbled with grey, look interesting when they are planted with dark green holly or box. If you intend to include a conifer, make sure you choose one which will stand being clipped – *Thuja plicata*, a fast, strong grower, for example. The variety *T.p.* 'Atrovirens' has bright green foliage.

I am envious of some reliably hardy, broad-leaved evergreens available only to my east-coast American gardening friends. When I was there on a wet November visit, the Meserve Holly Collection kept cropping up in conversation. I saw impressive specimens from this collection in nurseries and in gardens designed by that remarkable father and son team, Gale Nurseries of Gwynned, Pennsylvania. It struck me then that these particular hollies could make welcome and most unusual additions to the colour range normally associated with tapestry hedges.

Kathleen Meserve has managed to achieve what some people consider to be the most important plant breeding breakthrough of the 20th century. By combining the beautiful with the hardy, she has created magnificent hollies. The now famous *Ilex* 'Blue Maid' and *I.* 'Blue Stallion' are perhaps even outdone by the newest China hollies, *I.* 'China Girl' and *I.* 'China Boy'. *I.* 'Blue Princess' and berried *I.* 'China Girl' are both wonderful for hedges. *I.* 'Dragon Lady' has an outstanding, narrow pyramidal profile.

If you are a holly addict, as I am, Meserve varieties are wonderful, but not all, alas, are available in Britain. Even the holly sceptic, I am certain, will be converted by Mrs Meserve's progeny. Their performance has been consistently hardy, beautiful and berrying. They can be used in a tapestry hedge as a stunning and unusual component, in the landscape as specimens, or in the garden as reliable single hedges.

If you wish to lighten the overall effect with golden-leafed shrubs, my choice would be *Lonicera nitida* 'Baggesen's Gold' and *Ligustrum ovalifolium* 'Aureum'. Both are fast-growing and need only gentle clipping back halfway through the summer. If you want grey, then use either *Hippophaë rhamnoides*, the sea buckthorn, or *Cupressus glabra* 'Conica'.

Planning your planting needs care. For a tapestry hedge, a vertical plan is as important as a ground plan. You must take into account that the various shrubs need to grow to the same height and into each other; they need to be planted so that they combine, not compete, creating an overall effect of changing textures and colours that look right from a distance. A tapestry hedge must look well groomed. If you use too many different shrubs, you will not be able to make it look tidy. There should be a good section of each species, at least 90cm (3ft) wide, with individual specimens of beech and hornbeam planted 30 cm (12 in) apart, holly, prunus and osmanthus more widely spaced with approximately 60 cm (24 in) between each.

After your hedge has established itself, you may get gaps at the base; that is the moment to fill it out with *Buxus sempervirens*, a hardy standby which thrives even in rather starved conditions.

One more thought before we leave tapestry hedges – they need a well-prepared ground and constant maintenance. Do not plant yours and then think, 'That's that, except for an annual haircut.' If one of the shrubs is not competing as well as the others, feed it with fertilizer or manure.

The subtle interwoven quality of the yew, box and holly tapestry hedge at Rodmarton Manor is the result of years of careful maintenance and gentle persuasion. Holly has been allowed to grow into cones, 60 cm (24 in) high, at intervals out of the hedge top, lending both incident and elegance to the solid mass of evergreen foliage.

WINTER PICTURES

DESIGNING a new garden or redesigning an old one, you are not likely to have planning for winter in the forefront of your mind. During the spring, summer and autumn, you will have been visiting gardens and noting the plants and plantings you most admired and coveted. Your notebooks will be filled with the names of roses, trees for spring blossom and autumn colour, bulbs and summer flowers – specially loved plants, plants which will clothe the garden from March to November.

Quite different criteria determine the composition of a specific planting when the aim is to make the garden look appealing in winter. Then, herbaceous plants will be below ground and deciduous trees leafless. Three months and more, nature's exuberance will be muted, so it is doubly important to have chosen plants that stand up to the starkness of naked trees and bare earth. To have made notes is important; better still to have taken photographs so you can improve on what you had last year.

I like to imagine my garden as a series of pictures, each one merging into the next and leading on easily from the last.

Views from the House

Looking from your windows is an important part of gardening life in winter – however hardy you are and however keen to be outside, rain, frost and snow make working the soil difficult, often impossible. 'Whatever the weather, there is a garden out there,' Allen Paterson, director of the Royal Botanic Garden in Hamilton, Ontario, has remarked. 'It may not be possible to

The view from an upstairs window at Barnsley is as compelling in winter as it is colourful in summer. The black sentinel yews cast long, dramatic shadows across the snow-covered lawn towards the herb garden. The low clipped box hedges that define the herb garden make a bold winter statement; in summer the shape of the box diamonds becomes less distinctive as the herbs grow taller.

work in it (you may not want to), but windows give on to it and it is obscurantist not to enjoy the visual pleasures that exist.'

The three windows you look from most are your bedroom, sitting room and kitchen – you must think about each view carefully. I am lucky enough to have my bedroom overlooking a main axis in the garden, where fastigiate yews flank a stone path. The path goes on through an iron gate in the Cotswold stone wall, and leads my eye beyond to a silver birch with an especially white bark. There are some winter days when the sky is dark with storm clouds and then suddenly the sun bursts through, its rays high-lighting this almost ghostlike tree. Such effects would be nothing like as striking if they happened too often; it is the rarity of the moment that counts. Every morning my path gives me a feeling of anticipation – I must get up, go outside and enjoy the garden, and maybe escape through the gate into the world beyond.

David Hicks has created in his Oxfordshire garden the feeling that the view goes on and on. He has made an opening in the belt of trees which bounds the field beyond his garden – an idea evolved by Le Nôtre when he was making his great creations, first for Fouquet and then for the Sun King at Versailles in the 17th century. If you are lucky enough to own the land beyond your garden, think of planting trees out there with attractive winter bark or shape, in fact an echo of my silver birch. In my dreams I long to plant an avenue of trees across the first field at Barnsley – from a farming point of view, I fear it would be sadly impractical.

If your garden is surrounded by a wall, a fence or a line of trees, remember that while your view from downstairs may be restricted, from upstairs you will be able to see beyond the boundary. Your planting will depend on what lies beyond – do you want to hide it or to enjoy it? If you have a stunning view over countryside, do not distract the eye from it by fussiness in the foreground. If unfortunately for you, your distant outlook is unattractive, then concentrate on creating interest in the foreground. If there is a woodland or shrubbery, then create a definite path, straight or winding, to entice you on. If the ground slopes away, you will look at the tops of your shrubs and your trees will appear shorter. If the land rises, the opposite will happen.

If your bedroom window looks down on to a terrace, as well it may, then remember that you will be sitting on that terrace during the summer and seeing it throughout the year. This is where I like to have plenty of tubs, as well as urns and alpine troughs. I discuss the planting of these on pages 62-5. Keep the warmest corner free of plants and pots, however, so that you have generous space to sit during moments of winter sun.

The plants you choose to grow around the terrace will probably be mostly spring- and summer-flowering, but you should have some evergreens, and winter scents are essential here, for those special sunny winter days when the weather seems to have forgotten its pattern of behaviour. A few from which to choose are *Daphne mezereum, D. blagayana*, winter viburnum, sarcococca and winter sweet.

Then there should be clumps of early bulbs coming through the paving. Combine these with low shrubs and plants which retain their leaves in winter – crocus with dianthus, lent lilies with rue, *Iris reticulata* with rosemary or rock roses. The brilliant yellow *Iris danfordiae* should be tucked into a corner. Snowdrops can push their way through carpeting thymes and even tumps of ivy. Christopher Lloyd wrote in a *Country Life* article that ivy leaves are a good background for snowdrops in a winter vase. I took note and have planted *Galanthus elwesii* through *Hedera helix* 'Luzii'; they make an excellent January combination. Ivy is a good edging for a formal area if you remember to keep it clipped and tidy. If you want a solid ribbon, you must use a small-leaved variety like *H.h.* 'Glacier' or *H.h.* 'Sagittifolia.'

Other evergreen shrubs suitable for a sheltered terrace beside the house are *Coronilla glauca*, dense and busy, the dark green, purple-stemmed *Hebe* 'Autumn Glory' and the purple-leaved *H.* 'Mrs Winder', *Sycopsis sinensis*, with leathery leaves and yellow February flowers, camellias in variety and, for a really warm spot, the creamy-flowered *Buddleja auriculata* and *Grevillea rosmarinifolia*, whose deep green leaves have pale undersides.

How do you end your terrace? It can just merge quietly into the garden, becoming lawn or border; or if you prefer a feeling of enclosure and a framework, you can define the division between terrace and garden by railings, a simple, low brick or stone wall, or a pretty, painted wooden picket fence.

Ideally, the view from your desk or sitting room window should be different – restful, not a pathway inviting you along it so that you neglect your work indoors. To me it is one of the most important views – you will probably sit here hours longer in winter than in summer – and so it should be carefully planned to change in character as the seasons advance.

If you are making a ground plan for the border you will see from your sitting room, bear in mind that this view in particular will be a still life framed by your window, since you will be seeing it usually from the same angle, whereas borders elsewhere in the garden become unfolding panoramas as you move along them.

Include shrubs for winter effect, then infill with herbaceous plants. Plan in hand, look from your viewing point in the house and visualize where you are going to place your shrubs. Go outside and put up bamboos of the appropriate height. If you find this difficult, get a few large objects such as a bucket and a wheelbarrow and put them in the border, covered with a garden sheet or sacks, to show where the evergreen shrubs are to go. A fork and spade can represent the deciduous shrubs like deutzias and peonies. This will help you to discover if your plan is right. Remember to look at it from every other angle, too, and to think about every season of the year. Time spent thinking about it is never wasted; it is important to get this initial planting right so that things do not have to be moved around later.

During the months when the herbaceous plants are underground, the borders further away must have been planted to provide enough incident – heights, massed contours and individual shapes. Height can be brought by standard deciduous trees, so long as they do not have too dense a foliage to cast heavy shade on the border in summer. *Gleditsia* 'Sunburst', with its branched spines, is ideal, and can be kept to the height you wish by clipping the branches back hard in spring.

For middling height, use summer-flowering shrubs such as tree peonies, brooms, deutzias, philadelphus and *Spiraea* x *vanhouttei*. Winter light will be colourful and effective on the bark of all these shrubs. Prune their dense hanging 'skirts' away and plant groups of early bulbs around them – crocuses, scillas, narcissi and grape hyacinths. Whole ribbons of crocuses can come up through the middle of this border, and later-growing herbaceous leaves will hide their dying foliage.

Then for contours there are the more solid evergreens – *Osmanthus burkwoodii*, cypresses and junipers – in horizontal and upright shapes. In winter, viewed from a distance, they provide strong rhythm and prevent the borders from looking flat and uninteresting.

In my garden, I get great pleasure in winter from seeing an evergreen *Juniperus* × *media* 'Pfitzeriana' growing beside *Santolina chamaecyparissus* and underplanted with *Stachys byzantina* (syn. *S. lanata*). We have clipped the lower branches off the juniper to lighten it and give it a more attractive shape. It looks far more elegant now, showing its lower branches a lovely dark red. In this border we allow some of the herbaceous plants to keep their flower stems until spring, especially *Phlomis russeliana* (syn. *P. viscosa*), bergamots and various grasses – their dried seed-heads look like skeletons when frost or snow is on them. The tits and finches enjoy themselves too, swinging on the stems and carefully,

In a small garden all views are telescoped; imaginative planting has to take precedence over grand gestures. Colour schemes create the vistas and encourage the eye to travel from one group to the next. In Nada Jennett's garden near Bristol in Avon, the summer look is lush, leafy and inward-looking, with the blues, greys and purples of flowers pushing the overall greenness into the background. In winter the lawn becomes a force to be reckoned with, the countryside beyond is revealed, and the evergreen shrubs become dominant in shape and colour within the garden.

picking out every available seed.

This far border may be backed by a wall, a hedge or a fence, any of which in winter will stand out in greater relief when plants in the border are low or underground. If you have an evergreen hedge, especially a tapestry hedge (page 45), you will have year-long interest.

On the other hand, a wall, as I am constantly being reminded by visitors envious of our 18th-century example, provides wonderful opportunities for climbers and shrubs which like a degree of protection. Choose some with striking winter foliage or flowers. Variegated ivies – *Hedera helix* 'Goldheart' and *H. colchica* 'Sulphur Heart' (syn. *H.c.* 'Paddy's Pride') – are good examples. The yellow in the two ivies may be echoed in one of the most reliable of all winter-flowering shrubs, *Jasminum nudiflorum*.

There is no reason why other shrubs which can be clipped should not make good wall patterns in winter. Holly, though initially slow-growing, soon gets into its stride, and I would like to experiment with *Ilex* x *altaclerensis* 'Golden King' and *I. aquifolium* 'Silver Queen'; clipped back hard to the wall, they would make an original and exciting impact, their variegated leaves showing up well on cold winter days. With enough space (2.75 m/9 ft) on a south-facing wall, I would like to grow a cultivar of *Prunus mume*, the Japanese apricot, for its fragrant winter flowers. Russell Page, one of the greatest of postwar gardeners, when he saw how well a *Fremontodendron californicum* was doing at Barnsley, remarked. 'Why don't you espalier this along the wall?' Why not indeed, I wonder, for the foliage is pretty, a few blooms are usually out at Christmas and it continues to flower spasmodically throughout the summer. One day perhaps I will find the right wall.

Years ago I was intrigued to read something Sir Thomas More wrote: 'As for rosemarie I lette it run all over my walls, not onlie because my bees love it but because it is the herb sacred to friendship, whence a sprig of it hath a dumb language.' If confirmation were needed, I learned later that the 16th-century German chronicler, Hentzner, had reported seeing rosemary growing against the walls at Hampton Court.

A herbaceous border made magical by early winter frost: the leaves of lady's mantle, Alchemilla mollis, *are outlined with rime and the heads of sedums have changed colour from brown to silver-grey. Evergreen box balls, barely evident in summer when herbaceous planting makes strong groups of vertical colour, in winter create a rhythm along the entire length of the border.*

So I tried it at Barnsley and very effective it has been. It is the only rosemary in the garden – apart from me – to have stood up to the ravages of the last few winters. The evergreen *Garrya elliptica*, with its winter tassels, can also be grown in this way.

Although it is slow to get started upwards, one of my favourite evergreen climbers is *Euonymus fortunei* 'Variegatus'. It is very effective with its small, shiny gold and silver leaves, which light up in late winter as the young leaves open. Any of these evergreen climbers can act as host plants for more delicate summer climbers – eccremocarpus, morning glory, sweet peas, maurandya. Each year you may vary them.

Now think about the view from your kitchen window. Many of us these days spend a lot of time in the kitchen, using it as a breakfast room and a place where the family gathers. Modern houses try to make the kitchen a sunny room, maybe with a large window. If so, you will be looking at the same view from your sink or table from the same angle day in, day out, and it will need the same careful thought given to its planting that you give to the view from your desk.

In old houses the kitchen rarely faced south, so if you live in an old house yours may well be on the east, or even the north, side. My sink faces north and we have done our best to lighten the small enclosed area it looks out onto with a rampant specimen of *Jasminum nudiflorum*. Its yellow sprays are lovely for flower posies – pick them in bud and they will soon open in the warmth indoors. Growing with it is the Rambler rose 'American Pillar', which manages to flower profusely in July, in spite of lack of sun.

Another winter-flowering shrub which prefers shade is the bush honeysuckle, *Lonicera fragrantissima*. It is often part of my Christmas table decoration, along with *Chimonanthus praecox*, which, however, likes to face the sun. One of my favourite evergreen shrubs for flowers to brighten January days is *Viburnum tinus*, but it does not like shade that is too dense. If your kitchen window wall itself is in shade and is large enough, *Hydrangea petiolaris* is a good choice; at the foot of the wall could be forms of *Camellia japonica* and *C. reticulata*. The variety 'Captain Rawes' has wonderful semi-double flowers,

From a window at Levens Hall in Cumbria, the view opens onto an impressive array of evergreen topiary shapes, their beautifully sculpted forms and the varying colours and textures of their foliage high-lighted by the winter sun. A narrow gravel path leads the eye beyond the multitude of shapes towards the open countryside beyond – the formal topiary standing in sharp contrast with its surroundings.

which open in February until early May. *Daphne mezereum*, an essential shrub for winter, does well in shade and will probably be in flower by February too.

Many evergreen groundcover plants also like shade – bergenias, hellebores, lamium, brunnera, *Mahonia aquifolium* with striking bronze-red winter leaves, *Pulmonaria officinalis* (sometimes in flower in January) and *Saxifraga* x *urbium* with pretty leaf rosettes. Most of the ivies will survive shade and keep their variegated colouring. As long as you are prepared to restrain them, the vincas are ideal cover; it is best to go over them with clippers in the winter. They will then make new growth and flower much more profusely in the spring. *Calluna vulgaris* in its many forms makes a good carpet of winter colour on poor acid soil. The foliage tints are best in winter and range from dark greens through bright green, and gentle to startling golds. The form *C. v.* 'Silver Queen' has fine silvery-grey leaves.

I think it essential to have a door leading from my kitchen directly into the garden. The first thing I do when I come down in the morning, winter or summer, is to open this door and have a sniff outside, to set me right for the day and to get a feel of the weather. This is the place for the herb garden – handy for cooking and nearby for scent.

The Approach

The approach to your house and your front garden is every bit as important as the views you have from inside the house. To make a wealth of colour is easy enough with spring bedding and summer exuberance, but the other months need careful planning. For a small driveway, I would recommend planting a basic framework of evergreen trees and shrubs, creating with single specimens or small groups a border of varied shapes, sizes, textures and shades to welcome you and your visitors. It is sensible to protect the border with a low wall to keep cars from trespassing on it.

Start with the low-spreading junipers – *Juniperus sabina* 'Tamariscifolia' and *J. horizontalis* 'Glauca' – in the front of the bed and also perhaps as a foil for the long trailing shoots of *Cotoneaster dammeri*, an evergreen with berries which reliably last well into the winter. Early last winter I was struck by a bank of this shrub growing along a much used pathway above a 90 cm (36 in) retaining wall at the Royal Botanic

Above: Cotoneaster horizontalis, *covered with red berries, has been carefully pruned to frame the windows of this stone facade, with neatly clipped* Lonicera nitida *making attractive bollards each side of the door.*
Right: The weeping willow, clipped box, viburnum in flower and clusters of snowdrops provide an abundance of interesting silhouettes and unexpected colour that make the approach to this cottage as compelling in winter as in summer, when the simple framework is disguised by a mass of flower and foliage colour.

Garden in Hamilton, Ontario. The director, Allen Paterson, assured me that it looks impressive throughout the cold winter.

Above the junipers you may well need some accents of height – candleflame shapes or pyramids. Conifers are best for this – particularly chamaecyparis, other junipers, thuja and yew. Beside an Irish juniper, *J. communis* 'Hibernica', try planting *Chamaecyparis pisifera* 'Filifera Aurea' given sufficient space, its golden, whiplike branches contrast well with the juniper's upright habit of growth. A scattering of conifers gives weight and permanence in winter, but you must not let your border be dominated by them. Before making a choice, learn about their rate of growth by looking at other people's gardens. Then go to an old-established nursery to find out what they cost, what is available, and, much more important, what heights they will reach. If you go to a garden centre where expert advice is not readily available, you ought at the very least to take an explanatory catalogue with you, or in fifteen years' time you may well find youself with a 4.5 m (15 ft) *Thuja plicata* dominating a small bed. (You will undoubtedly remember the kind helper at the garden centre who assured you that it would only grow as high as you.) Incidentally, this thuja is an excellent evergreen to plant as a screen between you and your neighbours, or as a backdrop to a tennis court. But remember, when it exceeds the height you can reach on steps with your long clippers, top it.

There are several broad-leafed evergreen shrubs which flower in winter, have attractive foliage and make a good foil to the conifers. The mahonias are often in flower as early as November. They have strongly defined pinnate leaves – these are prickly, so keep them away from a narrow pathway but not too far so as to miss their lily-of-the-valley scent. I would put them on the west side of the house, where they will be in partial shade and where you will be able to enjoy the fragrance of their flowers. I agree with the author Graham Stuart Thomas that 'the true *Mahonia japonica* is one of the most impressive of foliage plants'. Seeds of *M. lomariifolia* were brought back from China by Lawrence Johnston, the creator of Hidcote; a hybrid of this and *M. japonica*, named *M.* 'Charity', is one of the best. Others to choose from are *M.* 'Buckland' and *M.* 'Lionel Fortescue'. I believe every garden should have a specimen of mahonia – why not plant it here by the drive, where you will not mind sacrificing some racemes of flowers to bring indoors?

A short driveway could be the ideal place for *Hebe* 'Autumn Glory' and *H. rakaiensis*, both evergreens with small, glossy

leaves which make tidy 90 cm (36 in) mounds. If you plan a yellow theme, then the hollies *Ilex* x *altaclerensis* 'Golden King' and *I.* x *a.* 'Lawsoniana', and the golden yew, *Taxus baccata* 'Dovastonii Aurea' would show up well and can be kept in shape and size by clipping. For shade you might choose sarcococca, skimmia and *Lonicera pileata*.

Obviously you will want to leave space for summer-flowering shrubs as well, but you should think of including some which may be at their most appealing in summer but also look attractive in winter. The evergreen *Osmanthus delavayi* bears its scented flowers in spring, *O. heterophyllus* in autumn, but both have good dark evergreen leaves and a dense habit you will welcome in winter. *O.h.* 'Variegatus' is slower-growing but has strikingly variegated leaves.

Interplanted with the shrubs could be bergenias, epimediums, pulmonarias, *Arum italicum pictum* (whose leaves disappear in summer), dianthus with good grey foliage, and, for a bit of height, pampas grass and the New Zealand flax, *Phormium tenax*. For flowers, early blooming crocuses and *Iris reticulata* look well around the base of shrubs, and I love to see *Helleborus orientalis* planted high enough to see into their hanging faces.

If, on the other hand, your house stands well back from the road at the end of a drive – mine is 45 m (150 ft) long – you must adopt a different approach. Plant bold groups and keep the scheme simple, or you will find you have driven past individual plants before you have seen them. For decorative winter stems, a group of *Cornus alba* is a good choice; *C.a.* 'Kesselringii' has purple-black stems and *C.a.* 'Sibirica', the Westonbirt dogwood, bright crimson ones. Both look attractive in summer as well.

Berberis are vicious shrubs and need to be placed where they will not tangle with the lawn mower, but they are effective when grouped and there are several to select from for winter leaves. *Bupleurum fruticosum* is another shrub which needs careful positioning: in summer its flowers attract flies, so keep it away from the house – a long drive may be just the place for it. It will grow well on a windy site and has particularly attractive grey-green leaves.

Here you might also plant evergreen elaeagnus, which has shiny leaves useful for picking and is fairly fast-growing, providing shelter from both wind and noise. If you live by a busy road, one of the best shrubs is *Viburnum rhytidophyllum*. Fast-growing, it has largish leathery leaves which make an effective sound barrier.

I have lined my drive with winter-flowering bulbs – the golden winter aconite, *Eranthis hyemalis*, plenty of snowdrops, scillas and crocuses. I also like the January-flowering *Crocus flavus*, which is a good spreader. I put them all in bold drifts, not scattered here and there singly. The lime-green flowers of *Helleborus foetidus* are just the right colour for winter: on rather grey days, they look luminous and act as a vivid foil to their dark green leaves. Put them in drifts too and be on the look-out for seedlings each summer, so that you can multiply them into a carpet. There are spectacular drifts of *Crocus tommasinianus* and *Adonis amurensis* at Winterthur in Delaware, the pale purple of the crocus contrasting with the bright yellow of the adonis. They both come through in late February, and I would like to copy this idea in my own garden.

Eranthis hyemalis *thrive on limestone or alkaline soil. Here in Oxfordshire, they complement the honey-coloured stone of the house, making a carpet of gold against which the strong dark shapes of the yews stand out dramatically in the distance. Essentially a plant to leave undisturbed, they multiply naturally by dispersing their seeds. As the leaves of the aconites fade and disappear, cow parsley,* Anthriscus sylvestris, *takes over.*

If you do not have a drive, you will want to treat this part of the garden differently, making it blend with your house. An English cottage or an American clapboard house might have beds of lovely billowing flowers – roses, hollyhocks and scented nicotianas – for summer, interplanted for winter and early spring with bulbs and low foliage plants, such as sympthytums, epimediums, rue and evergreen sedums. I would use the first of the *Narcissus bulbocodium* and *N*. 'February Gold' with the true blue *Scilla siberica* and *Puschkinia scilloides*. Rosemary, sarcococca and *Viburnum farreri* 'Nanum' could be added to provide scent.

If you have a path leading directly from the road to your front door, it should have an interesting pattern. The material of the path should blend with the house. Bricks laid in herringbone or in a weaving pattern can be edged with more bricks laid on their sides. Alternatively a lavender edging can look trim, while aubrieta will spill out onto the path, and so will snow-on-the-mountain. Whatever you choose, remember you will be seeing it at close quarters, so attention to detail will be all-important. It will be a case of planting in small numbers rather than bold groups.

American clapboard houses may be like English cottages, girt about by fences and hedges, or they may sit in a sea of mown lawn, so the whole feeling is more open and the planting close up against the walls. Because the houses are painted such clear colours, the shrubs appear all the more dramatic. I love the simplicity of this kind of landscape and the hard edges of the boxwood and other evergreens in the forefront.

The approach to a town house is generally rather different – the path is either central or, more usually, to one side. The architecture of the house often calls for a formal, symmetrical response. Box edging, box balls and clipped hollies are the most likely horticultural echoes. If that is your challenge, then make sure you include a small piece of lawn – you will find the green of the grass so restful in winter. Large trees will be out of the question, unless you face up to the task of keeping them pruned to shape and size. Hawthorns, limes and hornbeams can all be trained along espaliers or into mop heads. Their bare tracery always looks exciting when they have an icing of snow or a covering of frost to show off their bone structure. Grown as short standards and then clipped into mop heads or cubes, the evergreen oak, *Quercus ilex*, could provide an unusual answer in the search for formality.

Another unusual idea would be to have single or double-tiered espaliered fruit trees – apples and pears – edging the

Left: American clapboard houses generally have the simplest of decorative front gardens divided from the road by an expanse of lawn or enclosed by a plain wooden fence, and here made festive by the arching boughs and mass of bright red berries of Pyracantha coccinea *'Lalandei' under snow.*

Below: The brightly coloured stems of Cornus alba *'Sibirica', the Westonbirt dogwood, is a fine plant to look at from your house, cheerful throughout the winter months. Best keep them pruned every year but not until spring plants come through to take the place of the dark scarlet stems of the dogwood.*

Imaginative clipping of a common plant, Ilex aquifolium, *has created a dramatic double-tiered effect at the approach to Shakespeare's house at Stratford-on-Avon. It is both appropriate to the period of the house and a skilful and authentic interpretation of 16th-century gardening styles.*

path to the front door of a town house. Alternatively, you might like to have them as a central feature in narrow beds, (not more than 46cm (18in) wide), arranged in the shape of a St Andrew's or a St George's cross. Where the beds intersect, the crowning glory could be an evergreen such as a spiral or pyramidal box, or a yew clipped as a column. The fruit trees could be underplanted with evergreen groundcover – ajuga, small-leaved ivy, ophiopogon, low heathers or London Pride. Keep to one plant or the design will look bitty. Avoid anything that will grow over 15cm (6in) tall, or it will interfere with the fruit trees.

This does not mean that a town house has to have a formal approach; a front garden should always reflect the style of the house and the mood you wish to create – formal, informal, classical, romantic. My instinct is to keep first impressions simple, leaving delights to be revealed later.

Tubs and Containers

Tubs and containers provide essential garden pictures near the house in winter. Often you see them stripped of their summer flowers, left bare until it is geranium and fuchsia time once more. Filling them for winter can be a pleasure, a stimulus to the imagination.

We replant ours twice a year, in October and late May. The plants used in October will be there for seven or nearly eight months, and as I see them constantly it is important that they should be attractive and interesting. I usually have the same basic plan; it is just the final embellishment which varies with every new planting.

The tubs are half barrels with a diameter of 60 cm (24 in) and a depth of 40 cm (16 in) – plenty of space, in fact, for lots of plants. Their summer and winter plantings are quite different in character. In summer, the tubs are exuberant all the time. In winter, there must be a nucleus of evergreen, but also a feeling of expectancy, and as the weeks advance an excitement as each new early bulb shows through. Soon they will be coming into flower to please me and attract the bees.

As a start we change the soil completely, using compost from our own compost heap mixed with peat and perlite (1/2 compost, 1/3 peat and 1/6 perlite) with slow-release fertilizer added. The base of the tubs are permanently lined with broken crocks. We half-fill the barrels with our mixture and have a wheelbarrow load waiting.

On my terrace there are four tubs, symmetrically arranged, so that they have to be either identical or two matching pairs. I choose standard evergreens as a central feature, 60 to 90 cm (2 to 3 ft) high, in each tub – this usually depends on what is available, and we keep them from year to year – clipped hollies, *Rhamnus alaternus* 'Argenteovariegata', the variegated form of the sea buckthorn, or the standard *Euonymus fortunei*. Our standard golden privet and golden lonicera will soon be good enough to come into the list of possibles. In general, the need is for handsome standards which will be pleasing to look at

Above left: The elegant branching habit and subtly variegated leaf of Rhamnus alaternus *'Argenteovariegata' make it a fine standard specimen and one that lends colour and interest to the terrace at Barnsley throughout the year.*

Above right: Winter sun encourages the yellow crocus to flower abundantly at the base of the narrow-leaved buddleja, 'Dartmoor', which is left unpruned until early spring – the long shoots prevent young growth from developing too early when there is still danger of frost.

during those seven months. Your choice will vary according to climate and to what you have anticipated using and have grown for this purpose.

Next, the bulbs. Plant these in layers, to flower in succession, remembering that they will push their way through each other. The bottom layer will be at least eighteen tulips, planted thickly round the evergreen. Sprinkle them with just enough soil to make a bed for a dozen narcissi, then more soil and a few – perhaps only eight – hyacinths. Then put as many crocuses as possible as your top layer, choosing the species varieties, as they are the earliest to flower. Daffodils and tulips do not qualify as winter flowers, so you must have enough evergreen plants to cover the soil in December and January before the spring bulbs flower: winter pansies, sedums, violets, vincas to hang down, ajugas, lamiums, anything that the bulbs can push through.

Another successful combination, particularly effective for Versailles tubs in total shade, is a central feature of standard euonymus surrounded with *Lonicera pileata*. This evergreen honeysuckle spreads out horizontally in a fan shape and eventually becomes too wide, but lonicera does not mind being pruned back. It is a permanent planting and will last until the lonicera loses its freshness, which may not happen for three or four years.

The steps into my sitting room have rather special plants in summer and it is always a challenge to decide what to put there in winter. Single evergreen specimens in clay pots provide an answer – clipped box balls or pyramids, or small bay trees. Camellias and *Viburnum tinus* will give you colour, *sarcococca* and *Daphne odora* scent. Choose any of these and when they outgrow their containers you can replant them in the garden.

The other place I like to keep pots well stocked with plants for winter is just outside my kitchen door. (The area is paved, so I cannot have beds.) There are two main tubs, half barrels which we dare not move for fear of their total disintegration. One has a fremontia, which most years somehow manages to have a few flowers in bloom at Christmas, and a luxuriant *Lonicera nitida* 'Baggesen's Gold' bushing out at its base. It is a totally trouble-free tub; all it asks is to be watered in summer and gently pruned to keep it in shape. On the other side of the

The house in Magnolia Plantation, South Carolina, the home of Drayton Hastie, defies winter as the garden marches up the steps and along the balcony in pots of holly and longleaf pine, Pinus palustris, *backed by red-leaved nandina.*

door is the ivy *Hedera helix* 'Buttercup' – in fact the theme for winter here is definitely yellow.

I like putting early flowering bulbs in pots; to bring them on early, they are stood in frames and taken out as flower buds begin to show. Early crocus and *Iris reticulata* are wonderful for this, and so are some daffodils – 'February Silver', 'February Gold' and the miniature 'Hoop Petticoat'. Try different ones each year, and when they have flowered for you in their pots, put them out into the garden, where they will surprise you the following spring.

Outdoor Rooms

Outdoor rooms – spaces enclosed by walls or hedges – are to gardeners what squares are to town planners. Napoleon called the Piazza San Marco in Venice 'the finest drawing room in Europe.' Here we are not thinking of anything so grand – more of the intimacy of Hidcote's unfolding sequence of rooms of different size and different design. Usually square or rectangular, they can be formal or informal, but essentially they should be restful. Even the smallest garden room adds a new dimension by creating a secret garden-within-a-garden. In winter their boundaries, dimensions and evergreen containing walls become all-important.

In Mrs Winthrop's garden at Hidcote, the frames which the golden hops climb up in summer give height and emphasis to the four corners, even when the hops are below ground. If I had a similar design in my garden, I would choose an evergreen or shrubby climber, *Jasminum nudiflorum* or *Clematis cirrhosa balearica*, to go up the frames, or simply use a pyramid-shaped tree, such as *Malus* 'Van Eseltine' or Prunus 'Amanogawa' as host to an autumn-flowering clematis. This planting will create enough incident in winter to encourage anyone to walk through and enjoy it.

Alternatively, the centrepiece can be lawn. I know of several such pure green spaces, where the light green of lawn is answered by the dark green of the yew. If you decide on a less austere treatment than green upon green, it is important to have a wide bed round the lawn along the wall or hedge; a narrow one will look mean and uninteresting unless it is only there to take care of your climbers.

What you must remember when filling garden rooms with plants is the need to change the choreography season by season. Evergreens keep the lines intact, deciduous plants change the character dramatically – in winter you will be able to see through their tracery. To that extent, the element of surprise in what lies beyond will be lost.

A small garden room surrounded by walls or hedges high enough to conceal the outside world should generally have detailed planting, well-defined beds and interesting path patterns. The outline of the beds and the pattern of the paths will make this a pleasure in winter.

Many successful gardens make full use of garden rooms, some more defined than others. Hidcote and Sissinghurst were the initiators of this school of garden design, with connecting allées or arches or gateways. Tintinhull in Somerset is a simpler, smaller modification of this theme. It is clearly divided into compartments, each complete in itself, the paths guiding you from one to the next.

Penny Hobhouse, the gardening author who lives at Tintinhull as custodian for the National Trust, has described it for me: 'We are lucky to have a garden structure which shows at its best in winter. Evergreen trees – ilex, yew and our great, if suffering, cedar – a pattern of yew hedges making separate garden compartments and low box edging to flower beds. Box domes line our axial path, which stretches down the length of the garden; from it all the lateral compartments are reached'.

The central axis – a patterned stone path leading from the house through the grass forecourt and on into a more closely planted area and finally into the pool garden – is highly successful. What is interesting about Tintinhull is that the planting along the axis becomes more important and more detailed the further each section is from the house. As a rule, planting becomes simpler as the house recedes into the distance.

I like the idea of herbaceous borders in compartments within an allée. This is most effective when the border is backed by a wall, 'buttressed' by yew or box planted at intervals. This way you can keep to special colour schemes for each compartment. Yew gives a darker background for summer

Sissinghurst in Kent is truly a garden for all seasons, as this winter view of the White Garden makes clear. Whereas in summer the abundance of fresh greens and whites is dazzling, in winter it is the architectural features of the garden that strike you. Intersecting paths stand out boldly against the warm tones of earth, tree trunks and clipped evergreens and combine with the mellow red brick of the house and walls.

plants. To make a contrast in winter, choose a few shrubs with either coloured stems or evergreen leaves; the wall at the back should have eye-catchers – *Azara microphylla, Coronilla glauca, Buddleja auriculata, Fatsia japonica* (in shade), *Grevillea rosmarinifolia, Jasminum nudiflorum, Atriplex halimus, Clematis cirrhosa*. Not all of them are strictly speaking wall shrubs, but the more tender among them will be grateful for the wall's protection.

In the garden at Canon, not far from Le Havre, is a series of square, walled enclosures, identical in size, four of them in line where once the fruit was grown. A pathway runs through the middle of each square compartment, and there are simple archways in the centre of the intervening walls. Each room has a different colour scheme in summer, and in winter this effective vista cuts right through from end to end, the pathway edged with trim green box. It is one of the prettiest scenes I can remember.

You could adapt this idea to a smaller garden by having borders each side of a gravel path and dividing the beds into sections with buttresses, so that each becomes a separate planting. Enough distance must be allowed between buttresses to prevent the overall effect from becoming too fussy or busy. These cubicles would be ideal for sweet-smelling winter shrubs, and for groups of hellebores, interspersed with herbaceous plants.

Personally, I do not want a separate winter-garden-within-a-garden. I prefer my garden to be a garden for all seasons. If you have space enough, however, and are a keen follower of Gertrude Jekyll, this is her description of one she admired at Westbrook, near Godalming in Surrey: 'Another slight turn ... reveals a solid double arch leading into an enclosed space about 10.5 m (35 ft) each way. It is the winter garden – a delightful invention! Walled on all sides, the walling not high enough to exclude the low winter sun, it is absolutely sheltered. Four beds are filled with heaths, daphne, *Rhododendron* 'Praecox' and a few other plants. These beds, in company

The Pierpont Morgan rose garden is one of the principal outdoor rooms in Roy Strong's garden. Chosen as much for their scent as for their form, roses thrive here, given shelter and protection by enclosing hedges and clipped buttresses of yew. An 18th-century urn is the central feature of the circular bed which, edged with santolina, is planted with tulips and nepeta. Surrounding L-shaped beds depend on a massing of grey foliage plants infilled with annuals in shades of white and pink.

with the surrounding borders and the well-planted wall joints, show a full clothing of plants and a fair proportion of bloom from November to April. The brick-paved paths are always dry, and a seat in a hooded recess is a veritable sun-trap'.

A Winter Corner

A winter corner should be a place, however small, where you can be sure of finding those special flowers that brave the weather, those leaves that keep their colour and those berries that hang on through winter days. And it should be tucked away so that you have a positive inducement to walk out of the house to enjoy your choice of winter flowers.

My winter corner is somewhere I can take my visitors on a sunny day when they ask what is in bloom in the garden. It would be dull to have to say, 'You can see it all from the window'. The standbys are the hellebores. As with much of my gardening, they came there by accident. Nancy Lindsay inspired me to think carefully about plants, to look at them and see their beauty. She had a good collection of hellebores and would give away great clumps of them. Unfortunately, towards the end of her life her garden suffered because she was unable to weed as she once had; the dreaded ground elder insinuated itself. When my large clumps of hellebores arrived, we washed off every vestige of soil to get rid of any offending ground elder roots. I was still not really satisfied, so we decided not to risk planting them in any border where the gout weed could spread; instead we put them all along the bottom of the 1770 north-west facing wall.

At the time the wall was comparatively empty of climbers; we have since covered it completely with shrubs, clematis and roses, which flower through the spring and summer, and the hellebores make splendid cover for the base of the plant. Being close to the wall, they tend to become very dry in summer, so we have to remember to feed them, ideally in spring after they have finished flowering.

The first of the hellebores to flower is the Christmas Rose, *Helleborus niger*. The buds push through the soil and can be relied upon to give us a few lovely pure white blooms to put beside the crib in church and to decorate the dining room table on Christmas Day. Give them some protection with a cloche or bell glass or even just a pane of glass, not because they need shielding from the weather but because they can become mud-splashed in heavy rain and hidden in the snow – the perfection of their white sepals spoils very easily. Once you have established a clump, leave it alone and mark its place in the border, so that no well-meaning weeder disturbs it. Each stem should have two flowers opening in succession. If you are going to pick them, wait until the second bud has opened, or this will flag and remain closed.

There are several forms of the Christmas Rose, some with larger flowers and some with a pink tinge to their sepals – it is these which form the beautiful 'cup' of the flower. The petals

Above: Helleborus foetidus *is a striking and unusual plant, with its dark green foliage and fountains of pale green flowers.*

Right: If you plant a corner of the garden with winter in mind, you will have an incentive to go outside whatever the weather. Hellebores look lovely as single specimens but even better when planted in groups. In Beth Chatto's garden near Colchester in Essex, a mass of pink Helleborus orientalis *looks wonderful growing beside the water, flowering simultaneously with early daffodils.*

are small and insignificant around the cluster of stamens. My most prized group is *H.n.* 'Louis Cobbett', with large, pink-flushed flowers, given to me several years ago by Tony Venison, garden editor of *Country Life*.

The Lenten Rose, *Helleborus orientalis*, comes next. In favourable winters it will start to send its flower stems up soon after the New Year and will continue to do so right through until spring. The flowers are marvellous – delicate and humble as they hang their heads on their 46 cm (18 in) stems, varying in colour from pure white, palest green and cream through lovely pinks and wine colours. Most treasured of all by the connoisseurs is the almost black-flowered hybrid. I love best those with spots or flakes, usually of deep purple, on their cream or pale purple sepals, which you discover when you turn up their faces. I have just glanced at *The Plantsman* catalogue written by Jim Archibald and Eric Smith in 1970, and cannot do better than quote what they have to say about hellebores: 'The aristocrats of winter-flowering plants. All have fine, ground-covering leaves and establish slowly but surely in any good soil, preferably in shade. They do well in limy soils and appreciate a mulch of well-rotted compost in spring and some bone meal in autumn. Never disturb them ... they do not show anything like the true size or colour of their flowers until they are re-established after transplanting.'

I agree with everything they say, especially the fact that hellebores improve with the years. Mine have. You must be patient. Save all the seedlings which appear under the parent plant. Leave those which have space to develop where they are, pot up any others and put them where you can keep an eye on them. As you pot them, dip them in a mixture containing benomyl to prevent damping off. Mrs Helen Ballard of Colwall, Malvern, in Worcestershire, who has raised and sells many truly superb forms and hybrids, recommends treating established plants regularly with this solution. Another specialist nursery which offers these lovely hellebores is Miss E. Strangman's Washfield Nurseries at Hawkhurst in Kent.

Beautiful in a different way but just as desirable are the hellebores which carry their flowers on stems made the previous year. These are *H. foetidus*, our native plant, given the unflattering name of the stinking hellebore, and *H. corsicus*. Both have charm and should have a special place in the garden. The flowers of *H. foetidus* are small, but when several plants are grouped together they become very striking, with their dark green foliage and shade and under deciduous trees. *H. corsicus* is a stately plant, long-lived, not at all

fussy about where it is put and flowering freely early in the year. I have a few growing at the base of my wall, but perhaps the best place for them is on the edge of a wood or in a corner of the garden. The evergreen leaves are attractive in themselves. They, too, seed quite freely and seedlings appear around them which you should save and pot. They take at least two years to flower and from then on the whole plant will develop and send up more flower spikes.

Other plants to enliven a winter corner with their flowers and scent are species of snowdrops, *Adonis amurensis* and *Hepatica* x *media* 'Ballard's Variety', crocus, *Anemone blanda*, *Chionodoxa luciliae.* For extra height and scent you could plant your *Mahonia japonica* here. The two daphnes which qualify for winter are *Daphne mezereum* in white or purple (rather a short-lived shrub so do keep any seedlings that may appear in the garden) and *D. odora* 'Aureomarginata'. Both are sweetly scented and *D.odora* will waft its scent for a long distance across the garden. All daphnes appreciate protection from cold winds.

This would be an appropriate place for Abeliophyllum *distichum*, one of the shrubs which is unusual in cultivation. I do not grow it yet but am planning to for its almond scent in February. It is not a climber but needs the protection of a wall or fence. At the end of my winter walk I have the willow *Salix daphnoides* 'Aglaia', growing as tall as a tree; on sunny February days the plum-coloured catkins look wonderful against a blue sky, and if the day is warm enough the worker bees will be busy enjoying their pollen, preparing to take it back to their hives.

Slightly apart from the house, my winter corner is a place where I can rely on seeing some flowers out from Christmas through to spring. It is one of my most considered and enjoyable winter pictures.

Above: The blue flowers of chionodoxa, each with a white centre and yellow eye, are remarkable for the brightness and clarity of form that they bring to the garden in early spring.

Left: It is worth venturing outside, however cold the weather, to see a carpet of Galanthus nivalis pleno, *the double form of snowdrop, bravely flowering in a corner of the garden. Do not disturb your snowdrops but wait until each clump becomes a tight mass, large enough to be divided into one hundred individual bulbs. Once separated, they must be replanted immediately.*

WINTER COLOUR

WINTER'S PALETTE is clear and spare, restrictive enough to curb the excesses of even the most daring gardeners. 'Colour, in gardening, as in painting', wrote Gertrude Jekyll, 'does not mean a garish or startling effect...it means the arrangement of colour with the deliberate intention of producing pictorial effect, whether by means of harmony or of contrast.' But if we are to follow Miss Jekyll's advice when developing a plan for winter colour, we have to learn to know our plants. We can do this only by observing, noting, trying and testing.

First and foremost must be visits to 'other men's flowers', talks with other gardeners, visits to nurseries, botanical gardens – looking at groupings and massings, month by month, season by season. Then bookish sessions with other people's thoughts – the great plantsmen and -women such as E.A. Bowles, Gertrude Jekyll, Vita Sackville-West, Christopher Lloyd, Margery Fish and countless others, who write with passion on the way they have arranged their plants.

We do not have anything approaching that amazing book by the American writer Mabel Cabot Sedgwick, *The Garden Month by Month*. In 491 pages she records each month's flowers by colour and gives detailed information on their cultivation and height. To each plant she gives a number, linked to a colour chart printed at the back of the book. Instead, we have to make do with nurserymen's catalogues and illustrated dictionaries. But, in the end, it is our eyes, our colour sense and our trials and tribulations that shape the final picture of our garden.

Winter colour is nature's most sophisticated palette – a range dominated by subtle tones, sombre contrasts and striking highlights. As the rich shades of autumn give way to gentler winter hues, it is as though a hand has bleached the canvas. Perhaps the most dramatic change comes with the loss of deciduous leaves–horizons extend and foregrounds diminish as surfaces all over the garden emerge from summer seclusion and concealment. Many of the trees and shrubs appear skeletal after the lushness of their summer growth and yet it is these newly shorn textures and stripped torsoes that become the background colour of the garden. Whereas before it was the flaming reds and burning golds, now it is the gentler fawns and purples and the multitudinous browns and greens that dominate. For the winter gardener, the challenge is to enhance and build upon this mellow array.

The countryside in winter is quiet, the light clear and the landscape uncluttered. The combinations and gradations of colour in a garden should reflect these months of repose, this feeling of inner calm. Whereas in summer you will want to juxtapose bold colours and textures, in winter you would be wise to think of creating a muted theme, sometimes using a range within a single colour. Bright blues will have vanished from the garden, until the first scillas appear. Yellows, whites and greys, purples and reds are set now against the deeper shades of green and brown. Flashes of pure, bright colour, artfully placed, come as both a shock and a joy — brightly coloured berries and branches look striking against yews and hollies. When the light is strong, the white trunks of birch and eucalyptus are dramatic standing in the distance beyond an accent of solid green.

The way winter redefines gardens spatially, extending or contracting summer boundaries, is best described by what happens in my own garden. In one direction I look out from my windows onto lawn and herbaceous borders with the old Cotswold stone wall in the distance, scarcely visible in summer because the bright border colours are so dominant. Once a week, when the lawn has been mown, the pattern of the stripes, pale and darker green, stands out. In winter, gone are the lawn stripes, most herbaceous plants have been cut, and all that is left are the nut-brown stems of the tree peonies and the

Vibrant flashes of colour come as a suprise in winter's restricted palette. In Princess Sturdza's garden at Le Vasterival near Dieppe, a network of the brilliant red stems of Cornus alba *'Sibirica' stands out vividly against a background of evergreen, complemented by the reds of nearby hips and berries. A few of the bare red stems cut for the house will make a lasting display.*

compost-covered soil. I can look beyond to the wall, with its wonderful lichen and moss mottling the stone, golden and green, and it seems closer, protecting me from the outside world. There are just enough evergreen climbers, strategically positioned, each with a different density and colour, to prevent any feeling of bareness – a large-leaved ivy, *Hedera colchica* 'Sulphur Heart' (syn. *H.c.* 'Paddy's Pride'), blotched green on gold, a glossy *Itea ilicifolia* and, until the winter winds were too much for it, a matt green *Garrya elliptica*.

The other, more dramatic, example of the radical change in spatial dimension is to be seen from a west-facing window. Here, 36m (120ft) away, there are trees – sorbus, prunus and malus – growing with deciduous shrubs in rough mown grass. In summer they form a solid screen of differing greens. They constitute the wilderness or wild garden, completely apart, cool and shady and giving no clue as to what lies beyond. In winter the evergreens that were merely one green among many are now the only green among leafless neighbours. Effectively, they become the focus for the eye, foreshortening the view. The deciduous trees, shrubs and hedgerows which in summer merge and overlap, in winter allow the eye to travel through and beyond them, particularly when there is snow around and garden and landscape are reunited. As I look past their trunks, I have a sense of greater space. Each tree or shrub stands alone, a personality in its own right, with its own clearly defined silhouette.

With the exception of a cedar and a wellingtonia, an evergreen Portugal laurel and some hollies, every leaf on every tree has dropped by November and throughout the winter I am left with a wonderful see-through medley of brown and grey trunks and branches – some with berries still clinging where the birds have allowed them to remain, others with fat red buds waiting for spring – and the ghostly white bloom on the stems of *Rubus cockburnianus*. As the backdrop I see a beech hedge, 2.75m (9ft) high. This now defines the boundary of my garden, but with its shrinking leaves it has become sufficiently transparent for me to see the black-and-white cows grazing in the field beyond.

Over the seasons, nature presents us with so many different colours. Each plant's colouring is unique – now soft, now harsh, now shiny, now strong, now striped, now mottled. Faced with such a cornucopia, how is a gardener ever to plan a garden?

David Hicks offers an analogy which at least makes it sound possible: 'Getting colour right in the garden is very similar to a good sense of colour in interior design. A room which has no colour theme or colour scheme is a mess: a garden which has not had serious thought about colour is equally unattractive. It is a question of training your eye and seeing how other people's colour combinations please or displease you.'

This chapter aims to help you to 'get colour right' in your garden in winter. I have chosen to group the background colours – brown and green – according to their nature, and the stronger colours – yellow, white and grey, red and purple – according to their intensity and purity, starting with the lightest and brightest and moving on to the darker shades. But in the garden you can never be dogmatic. Colour is never constant: it changes, according to the light, the age of the plant, the weather – so many things.

Brown

Brown is a colour we tend to think of as autumnal. But the brown of autumn is flamboyant, whereas winter's browns are subtle and full of surprises, now rich, now gentle. Along with green it is winter's most recognizable mantle, nature's second cloak. Unlike green, it is almost entirely a background colour. Apart from barks and stems, which you should take time to look at for their own qualities, brown pervades the garden at every level, yet in an unobtrusive way.

Browns emerge in their different strengths once the green leaves of summer have fallen. I love the deep brown of bare earth, especially after we have just dug it and it looks clean and inviting. Then rain comes and darkens it still further. This is the time to mulch the borders, releasing yet more colour. Leaf-mould is one of the gardener's delights: it adds that special finishing touch, with its different browns of broken-down leaves and blackened horse chestnut shells.

When we have run out of our own leaf-mould, we spread rich brown Somerset sedge peat, with its appetizing texture. Many people use well-rotted cow and horse manure, which they spread liberally over the soil, particularly in the vegetable garden, before it is dug in for next season's pea and bean crops. It has a rich, chunky texture and encourages thoughts of summer plenty.

But brown is not simply a matter of earth or of mulch. There are also seeds and catkins: the clusters of alder and the varieties of willow catkin which open indoors from New Year

onwards. The best winter catkin-tree is the Turkish hazel, *Corylus colurna*, a tall pyramid whose long catkins often coincide with a full fall of snow.

Among my favourite browns are the common teasel, the herbaceous *Phlomis russeliana* (syn. *P. viscosa*), and many of the lovely ornamental grasses, which by winter have turned pale fawn. In my borders the gently browning spikes of *Acanthus mollis* – with their shiny green seeds and greying bracts which contrive to produce an overall effect of subtle, *haute couture* purple – stand out distinctively. So do the tall, mushroom-coloured stems of *Macleaya cordata*. *Crambe cordifolia* will have lost its seeds, but the brittle skeletons of its

At Great Dixter in Sussex, Christopher Lloyd has inherited and improved a garden whose winter colours are as rich and varied as a pheasant's plumage. Dried grasses, dead heads and leafless twigs present a wonderful array of beiges, fawns and browns, reminiscent of autumn colour but here designed to last throughout the winter. Clearly defined against the solid evergreen mass, the fine wispy grass stems are a magnificent sight.

Above: Tree bark is one of the most exciting winter sights. There are no leaves to distract, and sunlight and snow combine to bring out the rich colour of the trunks. Prunus serrula *has beautifully polished brown bark, with regular horizontal striations. If this were my tree, I would remove the ivy to reveal the full glory of its warm mahogany-coloured trunk.*

Right: The subtlety of winter's colours is matched by the gentleness of its light. At Kerdalo, Prince Wolkonsky's garden in Brittany, standing high above the château, a group of arching phormiums contrasts with upright conifers. The massed colours of these phormiums – pink 'Maori Sunrise', deep purple 'Dark Delight', green margined with pink 'Williamsii', and green margined with orange 'Sundowner' – are to me every bit as memorable as a border bursting with summer flowers.

stems are an elegant stone colour. Bring one indoors and use it as an original stand on which to hang your Christmas cards. Many of the eryngiums have decorative seed heads with a rosette of straw-coloured bracts, and of course *Sedum spectabile*, with their stiff stems and flat heads, play a part until the New Year. Nor are all the attractions at ground level – as you look upwards, the spent heads of the buddleja spikes are dramatic, silhouetted against the blue winter sky.

Barks and stems are with us all through the year, but it is during the winter months that we notice and appreciate them. Most gardens will have space for only one or two trees, the bark of which is their principal contribution in winter. They should go where the trunks will be seen clearly from the house or at a vantage point where they may be thoroughly enjoyed.

First there is the glory of the deciduous trees, stripped of their leaves with only their bark to speak for them. It is amazing how different the colouring of each trunk is. Anyone trying to decide which trees to plant should visit an arboretum when the leaves are off the trees – as well as when they are in full foliage – before making their final choice. It is worth all the trouble: trees may take years to grow but they give the most lasting pleasure.

Outside our front door is a gnarled old specimen of *Robinia pseudoacacia*. Its trunk is now 3m (10ft) in girth with incredibly beautiful furrows. On dry days the bark seems to be coloured in shades of grey and green, but when it becomes wet with rain it appears streaked with brown and red. It is a tree you have to plant for the next generation to enjoy and is one feature of the garden I would least like to lose. The London plane, *Platanus* x *acerifolia*, is also a tree for several decades ahead, but it is so handsome and has such interesting peeling bark that you should not dismiss it just because it is a slow grower. Walking through a garden or a grove of trees, the sight of the trunk of a single *Prunus serrula*, with its shining mahogany-coloured bark, will always take your breath away: caress it with your bare hands every day and it becomes as polished as the dining-room table. I sometimes wonder why some nurseryman does not graft a cherry such as 'Tai Haku', the 'great white cherry', onto the 1.5 to 1.8m (5 to 6ft) stems of *P. serrula*: you would then have magnificent blossom as well as a starting trunk. Too expensive, maybe.

For the smaller garden, one of the best trees for winter bark is *Acer griseum*, the paperbark maple, the mahogany-brown skin of which peels off, revealing new cinnamon-coloured bark. If you have a larger garden, with room for a number of

With the sun shining through its russet-brown leaves, the beech hedge in Nicholas Ridley's garden beside the River Windrush in Gloucestershire illustrates the owner's gardening philosophy: 'The charms of the garden in winter have to be achieved consciously: the design, the architecture and the winter colours are just as important as the planting of summer flowers ...'

trees, it is often better to group those with striking bark. The cumulative effect is more telling than any individual specimen standing bravely on its own, unless it is of great age and size. Three small trees with contrasting but complementary barks are *Arbutus andrachne* for cinnamon-brown, its hybrid *A. x andrachnoides* for cinnamon-red and *Sorbus aucuparia* 'Beissneri' for coral-red. Of these, the first is tender when young and the others hardy; I can guarantee you will enjoy them planted together.

I have vivid memories of the flaking, cinnamon-coloured bark of *Myrtus luma* (now known as *Luma apiculata*) growing in Cornish gardens, and much regret that it would not survive in my harsher climate. A possible substitute is *Betula albosinensis septentrionalis*. This birch is exceptionally hardy and is conspicuous for its orange-brown to orange-grey trunk.

For the smaller garden, there are just as many deciduous shrubs with exciting coloured stems as there are trees with striking bark. They all assume an important and different role in winter, the more vivid reds and yellows especially. They can be integrated into mixed borders, stood alone against dark backgrounds or, in the case of bamboos, formed into a solid screen between you and your neighbour or between one part of your garden and another.

Some shrubs can be planted especially for their stems, to stand out as startling whips of colour in the winter air. Most will have been chosen for their leaves and summer flowers; their stems, in shades of brown and grey, may not be distinctive in winter but they will blend in well with your winter planting. There are others midway between the two: although not brilliantly coloured, they have striking brown stems worth a place for this quality alone.

Stephanandra incisa has thin, zig-zag stems, while those of *S. tanakae* are elegantly arched and a rich brown. The leaves of the stag's-horn sumach, *Rhus typhina*, shine out for a week or two in autumn with brilliant colour, but its brown-felted stems are worth cultivating for winter; the female plant has the added charm of its cones of dark crimson, bristly fruit.

I must end my choice of browns on a high note, with two of the most beautiful hedges that are not evergreen – the beech and the hornbeam, *Fagus sylvatica* and *Carpinus betulus*. Both of these retain their russet leaves – 'fox-brown', as Vita Sackville-West called them – through the winter, but the hedge becomes progressively less solid as the leaves wither. I love to see the sun shining through them and to listen to their rustle as the wind blows.

Green

Green is the backbone colour of winter. It is too fundamental and comes in too many different forms not to be at the forefront of your planning for winter colour. Green has a thousand shades, from palest lime to darkest yew. It will reflect, it will absorb, but above all it will be there, year in, year out, the dominant colour of winter.

So I always feel sorry for those – and there are many – who think a garden can be a garden without grass. Grass is restful to the eye, soft to walk on, individual in scent and reliable in

Beckley Park in Oxfordshire is a small garden with a marvellous inheritance of clipped topiary. To walk down paths lined by cool, dark green architectural shapes, with no other colour in sight, is to realize that green is indeed the backbone colour of the garden, especially in winter. Do not be discouraged by the age of these clipped forms – you will be surprised how quickly box and yew can take shape.

colour – especially in our wet English climate. In America, garden lawns are more of a struggle, with drought a recurrent threat, but in winter they repay any effort – they frame borders, provide the permanent setting for planting, and spell out the base colour of winter.

Dark green conifers with massive, brooding outlines are the embodiment of winter's greenness: they are the first evergreen trees to come to mind. You have to look at them and decide for yourself which will fit your own garden best – merely to read about them is not good enough. You have to decide which you want as focal points, which as height accents, which as background or hedge. This is as important as deciding which colour you want – conifers range from blue-green to deepest green-black. Remember that most conifers have thick foliage right to the ground – planted close together, they quickly become a living screen. The taller they grow, the more they will confine you, hedge you in. If you live in a town, that is perhaps what you want; in the country, unless you are dogged by an eyesore, you want to see beyond your garden.

The nurseryman Alan Bloom has rightly described the role of conifers as 'the continuity links in a garden. It's a vital function, for they change but little in colour from one season to another, and in the darkest depths of winter the upright forms stand out like sentinels.' So which conifers shall be the first of the many greens a winter garden so desperately needs? To start with, there are the hedges, the gardener's ally in shaping the framework of a garden, the background of the planting plan. Best of all is the common yew, *Taxus baccata*. Darkest of dark greens, it is not as slow-growing as many people imagine. Once established, it will grow at least 23 cm (9 in) a year. As an alternative, you can use *Thuja plicata* 'Atrovirens', which has a more lustrous green foliage, but is never as solid or as reliable as yew. For tall evergreen windbreaks and backcloths, nothing is better than the western red cedar, *Thuja plicata*, its parent species. The lowest branches of old specimens hang gracefully and sweep the ground, putting down new roots which spring up as a skirt round the old tree. Crush its leaves and they smell of parsley.

Levens Hall in Cumbria is the ultimate giardino de verdura. Laid out in the late 1600s by the gardener to James II, the Frenchman Monsieur Beaumont, it remains a masterly example of the discipline clipped topiary imposes. The dense masses might be oppressive were it not for the rich variety of shapes and shades of green and the dramatic changes of scale.

If you want to use conifers as single trees for strategic positions in your garden, you must first decide which shade of green you require; then which shape is right; finally whether you need a form that will clip and can be kept within bounds.

Some conifers have a dense conical habit, like *Cupressus glabra* 'Pyramidalis', others a firm, upright form, such as *Juniperus communis* 'Hibernica'. The latter, used on its own, is ideal as an exclamation mark, punctuating sections of the garden. Then there are those which keep low, spread and can be used as groundcoverers, often as good at suppressing weeds as they are at providing winter greenery. They are particularly useful for concealing man-hole covers or other unsightly things you need to reach sometimes but want to hide. Spreading junipers are best for this task; *Juniperus procumbens* and its more compact form, *J.p.* 'Nana', are both quite low on the ground. For a slightly taller effect, but still predominantly horizontal, *J. sabina* 'Tamariscifolia' is a beautiful plant. It is excellent when sited as a single specimen at the top of steps, on a bank or as an infilling shrub in a shady border. The low junipers can be very attractive in winter when frost makes their leaves sparkle or when snow covers them. The tinge of grey coming from the reverse of their narrow leaves lightens their mass compared with sombre yews.

Two of the most memorable evergreen plantings to me are the dramatic stand of incense cedars, *Calocedrus decurrens*, at Westonbirt in Gloucestershire and the mature group of *Tsuga canadensis* 'Pendula' at Longwood Gardens in Pennsylvania, which are lovely throughout the winter. Their young shoots are almost luminous, with charactcristic whitc bands to the undersides of their leaves.

An interesting and proven use of conifers is to mass them together to makc a purely evergreen garden. Think of the 17th- century clipped topiary at Levens Hall in Cumbria, or the more modern examples at Beckley Park in Oxfordshire, and parts of the Ladew Gardens in Maryland.

Topiary on such a scale – although Beckley is quite small – will always be the exception. The more usual way of treating conifers is as occasional statements, adding strength to garden colours in winter. Their dark green is an effective counterpart to the more ephemeral trees and shrubs with alluring winter bark and foliage, and sets off the lighter and brighter stems of deciduous shrubs. In front of dark green yew, I have golden-stemmed willow and *Rubus cockburnianus,* its dark, purplish stems covered with silvery bloom. The red and the yellow stems of the dogwoods can be used in the same way.

If you use conifers in your mixed borders, they can be an effective foil for the early bulbs, for grey santolinas, and for the spikes of variegated iris leaves or handsome yuccas. In texture they can contrast with large, shiny bergenia leaves, with round, velvety-purple tellima leaves, or with the feathery quality of grasses. I have placed a deep green *Picea glauca albertiana* 'Conica' beside a clump of *Epimedium* x *rubrum*, the heart-shaped leaves of which take on such intriguing shades of red and brown in winter. The possibilities for combinations are endless.

Wall shrubs and climbers are halfway in height between tall trees and groundcoverers. If you choose, you can run through the whole spectrum of greens with the broad-leaved varieties. When the earth and tree branches are white with snow, climbers are at their most vivid. As the weather changes, they reflect its moods; as the sun falls on them they stand out, shining brightly.

First of all there are the ivies. In summer, the impact of their leaves is minimal, whereas in winter I love the intense glow of the common ivy, *Hedera helix*, on my Cotswold stone walls. It contrasts well with the yellow and grey lichen and the abundant soft moss, which, having survived the dryness of summer, comes to life again in the wet and damp. On rain-soaked days this combination can be wonderful – especially when the ivy is covered with black berries. Many of the ivies are strongly variegated and are discussed in later sections, under yellow, white and grey. Green ivy can be used to cover old tree stumps or to climb the bare trunks of trees. In a French garden, years ago, I saw ivy, clipped to about 46cm (18in), growing up the trunks of an avenue of lime trees – leg-warmers, in fact. The regularity of the pattern created by these gaiters was very effective. I saw them in late autumn, with a carpet of cyclamen joining each tree.

Among wall climbers the evergreen magnolias are the aristocrats. In much of England we have to settle for growing *Magnolia grandiflora, M. delavayi* and the sweet bay, *M. virginiana*, against the shelter of a wall. In Italy, the southern states of America and Britain's warmest counties, they make magnificent free-standing trees. Their leathery, shiny leaves give as much pleasure in winter as their fragrant, white-scented flowers in summer.

There are two winter-flowering evergreen clematis, both of which like shelter and a sunny wall. *Clematis armandii* is a vigorous climber with dark, attractive, hard-textured leaves and white flowers at the end of winter. It requires plenty of space

– you might use it to cover an unsightly roof or shed. *C. cirrhosa balearica* is more delicate, with fern-shaped leaves. Two climbing roses are also evergreen – the vigorous *Rosa banksiae*, with its fresh yellow-green leaves, and *R. laevigata*, the Cherokee rose, which by contrast has glossy, bright green leaves with red petioles.

Pileostegia viburnoides and x *Fatshedera lizei* are both notable for bold leaves; both need ample support and a warm spot. I love *Ficus pumila*, too – sadly not hardy in England as a general rule. Its small evergreen leaves will cover walls in frost-free places and become like wallpaper on pillars in a cool conservatory.

Itea ilicifolia is bushy but also needs a wall to give it support. Its long green tassels appear in late summer but persist until January. By then the silky catkins of *Garrya elliptica* – a shrub which enjoys an eastern or northerly aspect – will have stretched fully and opened. In spring and summer, when the flowers are over, these shrubs can support annual climbers, or even a clematis which needs hard pruning each year.

When you are planning a colour scheme for your wall, bear in mind the variety of shades and textures, and position your chosen plants so that green winter leaves and summer flowers complement each other. I plant sweet peas up *Itea ilicifolia* and *Clematis* x *jackmanii* through *Hedera colchica* 'Sulphur Heart' (syn. *H.c.* 'Paddy's Pride'), and you might try growing eccremocarpus through *Garrya elliptica*.

Shrubs are the mainstay of a garden. In summer they anchor the planting, preventing the exuberant annual flowers, herbaceous plants and wall climbers from getting out of hand. In winter the evergreen shrubs may well be the only plants above ground in a particular part of the garden, so their siting, the density of their foliage and the intensity of their colour, especially when the sun shines on them, will be all-important.

Hollies give me a feeling of security. Pruned to size or allowed to form their own shape, their glossy leaves reflect the sun. There are several green-leaved varieties to choose from. They stand out well against the darker evergreens, especially those with variegated leaves. And the density of colour which the smallness of their leaves produces makes their presence strongly felt in the garden; clipping them makes them firm in shape as well as solid in colour. The small-leaved Japanese holly, *Ilex crenata* 'Microphylla', is one of the hardiest evergreens and makes a dense shrub or handsome small tree. It can be pruned to make an excellent hedge, topiary, pleach, or espalier. As its name implies, it has small narrow leaves and I was surprised, when I first saw it, to discover that it was a holly and not a boxwood. (I have to remember when travelling in the States to call *Buxus* boxwood, not simply box.) The remarkably hardy *Ilex glabra* 'Densa', or inkberry, a native as far north as Nova Scotia and south to Florida, is another surprise, with its lax habit of growth (holly is usually stiff) and small, flat, dark green leaves. I have not found it listed in any of our British catalogues, but I am hoping to bring home some cuttings to propagate.

Two greens I am always grateful for in winter are the shiny texture of *Choisya ternata* and the leathery feel of *Osmanthus delavayi*. My first encounter with choisya, the Mexican orange blossom, was in a garden on the west coast of Scotland, growing beside a front door. It was in full flower and scent in January. I came home and planted one straight away. The osmanthus has different charms. It flowers in April, but you can watch the buds developing in winter and look forward to their scent when they open.

Evergreen berberis are useful either as hedges or as specimen shrubs. *Berberis gagnepainii lanceifolia* and *B. darwinii* are both excellent for either purpose. For groundcover the low-growing *B. candidula*, with small, dark green, glossy leaves, is extremely useful.

Pittosporum is another family I favour. It is not reliably hardy in my Cotswold winters but worth coddling slightly. *Pittosporum tenuifolium* has small, wavy-edged leaves with black stems. It is a good plant for picking, but take care not to spoil your bush. Evergreen shrubs are generally less hardy in their variegated forms, but the lovely evergreen *P.* 'Garnettii' I have found more hardy than *P. tenuifolium*. *P. tobira* can look wonderful as a hedge in warmer climates, particularly in the south of France and in California. I wish I could grow *P. eugenioides* outside, but it is unwise to tempt fortune with too many doubtfully tender plants; enjoy them instead in other people's gardens and stick to those which do well in your own climate and soil. I much prefer a handsome, if more ordinary, specimen to a struggling shrub chosen for its snob value.

The yuccas and phormiums, with their stiff, upright leaves, are both distinctive architectural plants, good in winter although best in summer. The yuccas come to us from Mexico and Central America and the flax from New Zealand. Gertrude Jekyll was especially interested in the use of yuccas, usually as final statements to mark the ends of a border. I like to position them so that you can see their stiff, erect leaves silhouetted against the sky. *Phormium tenax* has rigid, upright leaves, and

The dark green, glossy leaves of hedera, planted as wall or ground-cover, reflect light and provide interest at every season. The adult stage has the added attraction of small, greenish umbels, arranged in panicles, that contrast in colour and texture with the leaves.

like many architectural plants, looks well beside water; I prefer *P. cookianum*, with its gentler, more drooping leaves.

Groundcoverers and herbaceous plants, useful in the border and elsewhere in winter, can be used deliberately as a carpet through which spring bulbs push their way. I prefer on the whole to use the yellow- and grey-leaved forms, which are not too dense to look at, but there are several green-leaved ones I would not like to be without.

First, those universal groundcoverers, the ivies. I was delighted to find William Frederick extolling their usefulness. Of *Hedera helix* 'Baltica' he writes: 'This is the toughest, handsomest, most reliable evergreen ground cover I know ... The new growth is yellow-green and provides highlights to any planting ... It may be used by the design-minded gardener to create smooth, flowing lines.'

H.h. 'Sagittifolia' has pointed, shining green leaves and looks well growing in shade round the base of deciduous trees. I like to have snowdrops and also the lighter shades of *Iris reticulata* growing through the ivy, and I can see no reason why summer-flowering bulbs should not be planted with them as well to provide colour later in the year.

Under really shady trees where nothing else will grow, the large-leaved ivies will cover the ground. I have two special versions of *H. colchica*, the Persian ivy, for this purpose. One is the dark green *H.c.* 'Dentata', given me by the great gardener Peggy Munster, who had it growing in her famous garden at Bampton Manor in Oxfordshire, in a spectacular line under a sunless wall; we planted it under an important lime tree on our drive, where it thrives and looks good all winter. The other is *H.c.* 'Amurensis', another acquisition from that great plantswoman Nancy Lindsay; now, years later, it has formed an impressive dome 60 cm (24 in) high.

Three of my standbys for evergreen groundcover are scarcely mentioned in any of the books on this subject, except, of course, by Beth Chatto, and she praises them all. Beth has the most sensitive eye and imagination, combined with a great flair for plant associations, and I regard her with the utmost admiration for her contribution to English gardens. My chosen plants are *Pachyphragma macrophylla, Pachysandra terminalis* and *Waldsteinia ternata*. None of them is spectacular, but they all go on doing their own thing, covering the ground, keeping down the weeds, giving a show of flowers in due season and demanding no attention apart from a slight brush-up in spring.

The pachyphragma, which came to Britain from Armenia and the Caucasus, has round, shining green leaves, tinged with purple in winter, followed – just for a short while in early spring – by hosts of starry white flowers, before all reverts again to a green carpet. They are totally satisfying rather than spectacular. The rich green leaf of the pachysandra, with its wavy outline and shiny surface, is more exciting, but its white flowers are less striking. I have seen it, also in the variegated form, used to good effect in east coast American gardens. Both this and the dark-lobed waldsteinia will manage to exist in shade and even to cope with dry soil – they are happiest in an acid one.

Symphytum grandiflorum is as useful a weed suppressor as any plant, but do not make the mistake I did and put it in a border, where it outdid my treasures as well as the weeds. Its correct home is in the shrubbery or woodland; it will survive even under yew trees. Its leaves do die down in winter, but there is never a moment when some are not showing.

I would hate to be without the green vincas in winter. They will oblige in almost any shady corner assigned to them, and they are such useful plants for tubs, hanging over the edges and by February sending up flower spikes – blue, white, even plum-purple with double flowers. They will reward you, not only with constant greenery, but also with a patch which needs virtually no attention throughout the year. As with all good groundcoverers, you must not allow them to get out of hand.

A groundcover plant which is a feature all the year round is *Liriope muscari*, with narrow, dark green leaves. Like *Ophiopogon japonicus*, it is much used under trees in the warm states of North America, where lawns will not thrive because of heat and lack of rain.

Margery Fish and Gertrude Jekyll felt that the smooth, round, shiny leaves of low-growing bergenias looked best against stone or gravel. Grass is not their most telling background – I like to have them under or near conifers for the contrast of texture and leaf they provide. *Bergenia cordifolia* came to England in 1779 from Siberia, so should be absolutely hardy here.

Two eryngiums which make imposing green foliage plants for the winter border are natives of South America. *Eryngium agavifolium*, from the Argentine, has huge leaves of a dark blue-green, up to 1.5 m (5 ft) long and very narrow and spiny; *E. bromeliifolium*, whose leaves, narrow and finely toothed, make interesting rosettes, hails from Mexico.

Grasses may be used in different ways – as individual clumps in borders, merging in with herbaceous plants and

shrubs; as specimen features in the lawn; or as a whole bedful of several varieties. They appeal to many good gardeners. Christopher Lloyd and Beth Chatto both write appreciatively about them, and James van Sweden, the landscape architect in Washington DC, makes extensive use of them in his garden designs. He writes: 'Grasses change dramatically through the year. They move in the wind, make lovely sounds, and have soft colours and beautiful gold hues in winter. In addition to all that they provide food and resting places for birds.'

As he rightly says, many grasses are more golden or cream-coloured than they are green, but there is one smaller grass,

In summer it is far too easy to take green for granted, but in winter you become aware of the wide and varied range of tones and textures. Here, the dome shape of the broad-leaved evergreen pittosporum contrasts dramatically with the upright branches of the golden-green fastigiate yews, Taxus baccata *'Aurea'.*

Festuca glauca, with thread-like leaves only 15 cm (6 in) long, which is an attractive blue-green. It is quite suitable for the rock garden, where it looks better than in the border and will give a soft feeling to the stiffness of the 'rocks' in winter.

No true gardener's garden is complete without ferns. Whereas grasses appreciate sun, ferns require shade. Most of them will still be looking fresh in February; soon after that, I cut off the old fronds and await the uncurling of the new. But do not be too hasty in cutting down and tidying away. I love the fern fronds as they change colour from green to brown, some of them tilting over so that you can see their backs.

The deciduous ferns will linger on until set back by the hard frosts, but the evergreen ferns come into their own in winter. Their names are difficult to master, their beauty not to be ignored. The palish green Japanese shield fern, *Dryopteris erythrosora*, is about 60 cm (24 in); when you turn the fronds over you will find them decorated with bright scarlet spore capsules. *Cyrtomium falcatum*, the Japanese holly fern, has broad segments to its leaves which contrast well with more delicate foliage. It will make a bold patch on the edge of woodland, but is rather tender.

I would not be without the hart's tongue fern, *Phyllitis scolopendrium* – it is so useful in winter vases and will spring up from cracks in walls and paving. The tawny brown spores run in parallel lines, matching the colour of the stems. On some other ferns, the spores are in regimented circles the size of match heads. Another favourite is *Polystichum setiferum*, the soft shield fern, which will tolerate a drier soil than most. Turn over a frond and you will discover that the stems are swathed in almond-coloured papery scales.

Then there are the stems which show up especially green in winter. *Spartium junceum* is one; better still the stems of the flowering nutmeg, *Leycesteria formosa*, and of *Kerria japonica*. There is an unusual and most effective form of kerria, *K.j.* 'Kinkan', with a stem banded with green and gold, which I have only seen in Harold Epstein's garden in New York. For those gardeners who are keen on the unusual, it is certainly a shrub to be pursued.

When the light is right in my garden, I love to see the corky 'wings' on the young branches of *Euonymus alatus*. In spring and summer they do not draw attention to themselves, but by October I suddenly become aware of this spindle, and its interesting stems claim my attention through the winter.

If you were to pick a leaf or a stem from all the green plants I have mentioned in this section, you would be amazed at the kaleidoscope they offer. Like no other colour, nature's greens are all complementary: even if, with gay abandon, you used them all in your garden, you would still never wince at an unhappy clash.

David Hicks is unstinting in his advocacy: 'There is nothing more pleasing to me than all-green gardens: grass, yew, box, beech, hornbeam, ivy-covered walls, laurel... The plants within such a garden would be those that one grows entirely for textural interest, not for their flowers. I know certain gardeners who grow some plants exclusively for their leaves and texture, and abhor the flowers, nipping them off when they are about to bloom.'

I have visited all-green gardens in Europe and America where, far from being bored, I have been fascinated by the range of tones and textures they have to offer. One of these is Hidcote: after the assault on our senses of the marvellous range of its flowers, what a joy it is to return through the all-green enclosure of lawn and hedge, bank and mound as we leave the garden.

One of the most spectacular of all green gardens to be found is at Sir Harold Acton's villa, La Pietra, in Florence. He has described it to perfection as 'a series of broad terraces, levelled from the slope descending behind the house. The first is a long platform with a stone balustrade for statues at regular intervals, flanked by stairs on each side, which run down to the central terrace, enclosed by low walls and clipped hedges with niches for other statues. In the centre of this and the lowest terrace are fishpond fountains with circular basins, surrounded by stone benches and geometrical plots of grass contained by clipped hedges of box. A mossy staircase paved with coloured pebbles leads to the long grass allée between, with a colonnade roofed in by banksia roses above it on the right. Both terraces are planted mainly with evergreens. A peristyle of Corinthian columns separates the lowest terraces from the adjacent vineyard and a statue of Hercules stands vigorously in the centre with a couple of ancient cypresses behind him. Many paths running parallel with the hillside lead to stone arches and circular plots enclosed by hedges and statues. The whole garden is essentially green; other colours are episodic and incidental. Sunlight and shade are as carefully distributed as the fountains, terraces and statues. The wings of the little theatre are of clipped yew, with globed footlights of box. It is a garden for all seasons, independent of flowers.' To me La Pietra is a perfect stage set; its atmosphere dramatic in winter and full of surprises in summer.

Yellow

Yellow is the colour which lightens the greyest winter day: yellow berries surprise, yellow leaves glow, yellow flowers are a joy to bring indoors. It is too vital and vibrant a colour in the winter garden to be treated with profligacy. It is important to position your yellow shrubs and climbers, not together but strategically, to catch the winter sunlight: they look marvellous, too, as counterpoints to rich evergreen foliage. As Chris-

The fine feathery sprays of Juniperus x media *'Pfitzeriana Aurea', are green tipped with gold. Dew-laden spiders' webs hang from its spreading branches, emphasizing their light, airy quality. This golden juniper contrasts strongly and effectively with dark pyramids of Irish yew.*

topher Lloyd has pointed out, 'background is all important to this lively colour.'

Among the brightest yellows of winter are the fruits and berries. They do not have the solid impact of golden-leaved bushes, but they are scattered in vivid, welcome clusters along dark branches. We tend to think of berries as being red, so I like to have a few yellow ones to surprise and contrast. The small tree *Malus toringoides* has deeply cut leaves; by November they have fallen, leaving behind a mass of small fruits which hang on all winter and do not brown in the frost. Yellow to start with, they become bright orange as winter progresses; from a distance they look quite luminous and hang like a veil over the branches. *M.* 'Golden Hornet' has larger fruits, but unfortunately they brown in the frost.

Among the shrubs, the small, bright yellow fruits of *Cotoneaster* 'Rothschildianus' are there in profusion right through until January, when the birds settle on them. In a matter of days they are gone, but while they last they are a lovely sight with morning frost and sun shining through their branches and highlighting the berries.

I am old-fashioned and like traditional things. For me holly berries have to be red, but why not be original and try the yellow-berried form, *Ilex aquifolium bacciflava*? Graham Stuart Thomas will be your natural ally: he has written that its 'tint is far better than the red to crown the Christmas pudding, as it matches the rum and butter sauce'.

Yellow flowers on wall climbers play the same role as berries on trees and bushes – their bright studs of colour contrast with the solid shapes and heavy colour of the evergreens. *Acacia dealbata* has fragrant yellow flowers all winter long, and its lovely double pinnate leaves with their silver sheen are an added delight. It will survive on a south-facing wall in the south of England (my Cotswold climate is too harsh), and will flourish in an unheated conservatory. *Azara petiolaris* is another plant best confined to a conservatory, where it will flower more freely; its fragrant yellow flower buds open in January and February.

Fremontodendron californicum is hardier than both, and I feel justified in including it because, although its main flowering is in summer, I have been surprised by bright yellow flowers coming into bloom in January. By January, too, the branches of *Chimonanthus praecox*, the winter sweet, are usually covered in blossom; the fragrant, twinned yellow flowers often begin to open in December. Few English gardens are without a specimen of *Jasminum nudiflorum*, usually grown against a wall – it does not mind where it is grown and I have seen it cascading from a single post as a feature for height in the middle of a winter border.

It is exciting waiting for the buds on shrubs to swell and open. One day in late January or early February, you will suddenly become aware of a yellow haze over your *Cornus mas* – the individual flowers may not be special but when every branch on a large bush is covered, it looks marvellous. The flowers of *Corylopsis pauciflora* will also be opening in late winter. The dropping racemes of its pale yellow flowers have a delicious scent, so plant it where you can enjoy its fragrance.

The lovely winter-flowering hamamelis have flowers ranging from clear yellow to almost rich red. These witch hazels have been improved and increased recently at the Kalmtout Arboretum in Belgium by Robert de Belder. He and his wife like to accompany you and you will come away inspired by their knowledge and enthusiasm. *Hamamelis* x *intermedia* 'Jelena' is named for his wife. One of my most moving moments was at Winterthur in Delaware, when Hal Bruce, the curator, took me round. He has the best collection of witch hazels in America. They are growing together in a grove close to the main building, and I can guarantee that any visitor will enjoy the wonderful haze of yellows and browns created in the winter air. You see them from a slightly raised balcony walk in the main building, so when it is too cold there is no need even to go outside to enjoy them.

Yellow flowers are to be enjoyed for the rare and exhilarating accent they provide among the deeper colours of winter, but the greatest impact of yellow in a garden is certain to be foliage, particularly when it flows upwards. I have a special liking for the yellow-leaved ivies, *Hedera colchica* 'Sulphur Heart' (syn. *H.c.* 'Paddy's Pride') and the smaller-leaved *H. helix* 'Goldheart', but the climber I treasure above all others for its brightness on winter days is *Euonymus fortunei* 'Variegatus', with its brilliant yellow leaves. In my garden it has found an ideal companion, *Vitis coignetiae*, the Japanese crimson glory vine; together they clamber up an aged pear tree behind a stone bench furred with yellow-green lichen. Euonymus, usually thought of as a rather dumpy shrub, given the opportunity will quickly get going and cling happily to a wall or tree trunk. It has other virtues – it is not too demanding about soil, it will put up with a draughty corner, it provides good cutting material for winter vases – and, on clear winter days when the slanting rays of the sun catch the gold of its foliage, it makes a stunning picture.

A purely golden hedge is too startling for my taste, but if you decide to have one, *Cupressus macrocarpa* 'Goldcrest' would be a good choice. You must remember that clipping will remove the best colour; the young foliage is far yellower than the older leaves.

A shrub notable for its dense colour as well as being a real standby is *Lonicera nitida* 'Baggesen's Gold'. It can be allowed to grow unrestricted to form a bright, rather shaggy bush; I like to treat it in a more sophisticated way and keep it under control. I have it growing beside my kitchen door and on the terrace, too. In the severest frosts it will lose most of its leaves, but you need not be alarmed – it will come back into its own later with renewed vigour.

Elaeagnus pungens 'Maculata' is a shrub which seems governed by temperament rather than the weather: it has an unconscionable habit of dying, suddenly and unaccountably. It is worth your while to persevere, however, for its leathery leaves, heavily splashed with gold, catch the sun and make a fine display in winter.

To vary the heights of your yellow plants, try putting the golden privet, *Ligustrum ovalifolium* 'Aureum', with a group of contrasting shrubs in grass, where its golden shades will register more vividly. Alternatively, plant it in a border, where it will look attractive in winter and can be clipped back to any shape or size.

In your herb garden, the golden-leaved *Thymus vulgaris aureus* can be a glowing sight clinging round a box pyramid or as a foil to box edging, while *T.* 'Doone Valley' takes on an even brighter gold during the cold months than in the summer. The golden variegated sage, *Salvia officinalis* 'Icterina', is more unusual than the grey or purple varieties. Instead of planting the thymes and sages separately, you could combine them to create a tapestried carpet with gold, green and grey threads.

Yellow stems, especially planted in a group, can glow as vividly as yellow foliage. They may not be as eye-catching as those which strike red sparks in the winter light, but if *Salix alba vitellina*, the golden willow, is grown along a river bank with evening sun playing down or through its stems, it is every bit as dramatic as the scarlet variety. Some years ago, before I knew anything about gardening, seeing a stand of these I stopped my car in wonderment and asked the owners how the effect was achieved. Simple, they told me: just pollard them each spring, cutting the stems at a height to suit their site, and they will then send up wands of brilliant colour. My willow friends were right – it is simple to cut back to the old wood, but

At Barnsley an old pear tree acts as host for an unusually tall variety of Euonymus fortunei *'Variegatus', each leaf widely and irregularly margined with yellow – the effect is of light and soft, glistening colour. It is a pity that nurserymen seldom mention its climbing ability as, given support, this shrub will clamber to 6 m (20 ft) or more. Lovely in winter, it is even better in spring, when the young foliage predominates. In winter the bare stems of* Vitis coignetiae *are hidden by the euonymus.*

it takes a year or two to produce the bold effect I saw.

The weeping willow, S. 'Chrysocoma', usually planted for its graceful pendulous branches and its long-lasting golden-green leaves, has glowing yellow stems – barely noticeable in the summer but striking in winter.

There are yellow varieties of cornus, too. A group of C. stolonifera 'Flaviramea', the yellow-stemmed dogwood, looks wonderful if it is used with enough space around it. In a small garden, a single specimen can be effective, too, as a focal point.

Yellow grasses, ranging in colour from gold to straw, can make as much of an impact as yellow stems. Chief among

the small grasses is Bowles' golden grass, *Milium effusum aureum*, the soft yellow leaves of which are lovely throughout the winter. I have it growing under a golden form of cupressus and in company with golden feverfew, *Chrysanthemum parthenium aureum* (*Tanacetum parthenium aureum*, as it is now known). On sunny days, the whole picture is more like spring than winter.

Several of the grasses have feathery plumes and green leaves striped with gold, making slender splashes of gold and green in the border. Among the tallest is a new pampas from New Zealand, *Cortaderia selloana* 'Gold Band', with very narrow leaves. *Miscanthus sinensis* 'Zebrinus', an oriental grass, has gold and green stripes across its leaves, and *Carex morrowii* 'Variegata', the evergreen sedge, is worth trying in a damp spot in your garden. Plants like these with strongly variegated leaves – yellow and green or, more unusually, yellow and grey – appear lighter and less massive, but no less vivid, than those with more uniform colouring. Again, positioning is all-important. Variegated yellow shrubs could be planted against a dark background, or set off as a pair with contrasting markings; denser yellows could be planted in a distant corner where they will catch the eye from upstairs windows, as you come up the drive, or near a path leading to another part of the garden.

Because bamboos are dominant in shape and given to spreading, they should be used with discretion. One I favour is *Arundinaria viridi-striata* (now called *Pleioblastus viridi-striatus*), with yellow and green variegated leaves topping its 1.2 m (4 ft) purplish canes. A clump will make a firm accent, be useful in the wild garden, stop unwanted draughts round corners and act as a noise barrier.

There is a handful of low border plants which usefully retain their gold variegated leaves through the winter. The variegated strawberry, *Fragaria* x *ananassa* 'Variegata', a rampant spreader with many runners, lightens any corner all year round with its yellow and green leaves. *Tolmiea menziesii* 'Taff's Gold' has irregular green and yellow freckles on its round leaves. I have had it in a stone trough for several years, but I should really have given it a chance to spread at the front of a border. It is a good plant for the propagator because new plantlets form on top of the older leaves – hence its name, the piggyback plant. In winter we usually keep a small supply in pots in the unheated glasshouse as an insurance against loss and to bring indoors when flowers are scarce.

Some vincas have leaves stamped with gold. They are most

accommodating evergreen plants – you can use them in a variety of ways. Try, for instance, growing *Vinca major* through low deciduous shrubs (I have seen it with *Cotoneaster horizontalis*), or *V. minor* 'Aureo-variegata' trailing over the sides of tubs. Either form will make a dense carpet under trees or in a dark corner.

Among the variegated yellow shrubs, the hollies are so interesting and cheerful with their glossy leaves and enormous variety of markings. I grow quite a few, and would like to have more. On a recent visit to the southern states of America, I was impressed by the enthusiastic use made of many different hollies, and have resolved to be more adventurous myself. Perhaps my favourite is *Ilex* x *altaclerensis* 'Golden King'. Despite its name, it is a female, with almost spineless leaves brightly margined with gold. Clipped, it makes a wonderful bush – solid in shape, light in aspect. Another attractive form is *I.* x *a.* 'Lawsoniana'. To its dark green and golden leaf it adds a third colour – apple green. The mixture is unusual and eye-catching.

The golden variegated box, *Buxus sempervirens* 'Latifolia Maculata' (syn. *B.s.* 'Aurea'), does not make the same sort of impact as holly, especially when it is severely clipped in autumn – it is the young growth that is bright. If, on the other hand, you allow it to grow freely through the year, as an edging or a single specimen, it will stand out in winter.

A shrub I often use is the variegated snowberry, *Symphoricarpos orbiculatus* 'Aureovariegatus'. Its small leaves, margined with yellow, are refreshingly lightening in a border in summer and contrast well in texture with the heavier evergreens in winter. It is not until well into the New Year that its leaves fall.

I enjoy the way the leathery leaves of aucubas appear to be splashed with yellow paint – their markings vary so much. *Aucuba japonica* 'Crotonifolia' and *A.j.* 'Picturata' are both male, *A.j.* 'Gold Dust' and *A.j.* 'Variegata' both female: remember that the female will not bear berries unless you plant a male nearby.

Yellow is effective both in shade and in sun, but you must use the right plant in the right place. Some yellow forms of evergreen must have sufficient light and maximum sunshine to keep their leaves golden. These include the golden privet, *Hedera helix* 'Buttercup' and *Lonicera nitida* 'Baggesen's Gold', and even the variegated strawberry. If these plants are in partial shade, one side will be green, the other yellow. In a curious way this may look attractive, but it would probably not

Left: The large, golden flowers of the Chinese witch hazel, Hamamelis mollis, *have a feathery lightness that adds a delicate touch to the winter garden from December through to March. With its yellow leaves in autumn and the strong fragrance of its winter flowers, this witch hazel is a valuable shrub in all seasons.*

Below: Salix alba vitellina, *usually thought of as a shrub for a damp situation, is growing here in my own garden on shallow limestone soil. The sun shining on its yellow stems is one of the most pleasurable of winter sights. This is all the current year's growth – the long wands are cut back to the old wood in early spring.*

be what you had in mind when you planted it there. On the other hand, you can rely upon some of the ivies – *Hedera colchica* 'Sulphur Heart' and *H. helix* 'Goldheart' among them – as well as upon hollies, grasses and euonymus not to lose their golden markings, whatever their situation.

White

White and grey play much the same role as yellow in the garden in winter: soloists rather than orchestral players. In summer they may be massed to stunning effect – I am thinking of the famous White Garden at Sissinghurst in Kent – but in winter we do not need the breath of fresh air they bring by being grouped together. Instead, we need them as a foil for dark evergreens or to create incidental light among purple foliage or red berries.

The smallest, brightest whites of winter are its flowers. The aristocrats are the Christmas roses, *Helleborus niger*, the lovely blooms of the white *H. orientalis* appear later. Snow-drops provide the common touch and, unlike hellebores, which once established dislike being disturbed, increase of their own accord or can be divided immediately after they have flowered. It is so exciting watching them push their way up through the soil – with a little planning, you can have a succession from late autumn, with *Galanthus reginae-olgae*, until February, when the forms of *G. nivalis* come with a rush. I am lucky enough to have a wonderful inheritance of sheets of *Galanthus nivalis* 'Flore Pleno'. Another of my favourites is *G.* 'Samuel Arnott', whose rounded, bell-shaped flowers are, surprisingly, scented.

Slightly taller than snowdrops are the blooms of *Leucojum vernum*, the snowflake, which come through in February. Each flower, shaped like a minute Tiffany lamp, is tipped with green. *Iberis sempervirens* has intensely white flowers, closely massed, which can bloom from November until March. Its glossy evergreen leaves make a mound 46cm (18in) high.

For their scent, I enjoy the winter-flowering viburnums – *Viburnum farreri*, *V.* x *burkwoodii* and *V.* x *bodnantense*. *V. tinus*, although unscented, can be relied on for its flowers which last all through the winter. Few plants in winter have a scent like the *Sarcococca hookeriana digyna*, which really does waft upon the January air. In February come the scented white flowers of *Osmaronia cerasiformis* (now known as *Oemleria cerasiformis*), which hang delicately on leafless stems. *Ribes laurifolium*, evergreen and less well known than the pink-flowering currant, has unusual green-white flowers.

Among the climbers, *Clematis cirrhosa balearica*, which requires a south or west wall, opens its scented, creamy-white, bell-shaped flowers after Christmas, making an interesting contrast with its delicate leaves. Trees that have white flowers in winter include *Prunus subhirtella* 'Autumnalis' and *P. incisa* 'Praecox'. The former flowers as early as November, the latter in late winter. But quite outstanding among the white blooms of winter – so it must have a special mention – is *Magnolia heptapeta* (syn. *M. denudata*), the Yulan magnolia, with highly scented, brilliant white flowers shaped like cups. In a mild winter you will see its buds bursting open in February.

White berries may not be as striking as red, but they are still eye-catching and have the advantage of being less attractive to the birds. The translucent winter berries of the common mistletoe, *Viscum album*, hanging in clusters, are generally ignored by birds until well after Christmas.

The single, pearl-like fruits of *Sorbus cashmiriana* remain poised as if in mid-air long after their leaves have fallen. The white form of *S. hupehensis*, carrying its crop of berries in clusters, generally keeps its fruit until the severest frosts after Christmas. A few berries will probably be left on *S. vilmorinii*, too – small and round, starting off reddish-pink and gradually fading to white or cream. And the snowberry, *Symphoricarpos albus*, will frequently sport a good yield of pure white fruits on its bare stems from October to February.

White berries, even on the most fecund of trees, can never quite compete in winter with the shock of coming across the trunk of a tree that is white from top to bottom. If you study these trunks closely, you will find that their bark is made up of a range of colours, but the overall effect is almost pure white.

I thought I knew about the beauty of birch trunks until I saw them in the Morris Arboretum in Pennsylvania, where the white birch, *Betula pendula*, is as white as snow. Hugh

The White stems of Rubus cockburnianus *(syn. R. giraldianus)* *writhe in the winter air. Like the corkscrew hazel,* Corylus avellana *'Contorta', it is hard to believe that it is not the product of some laboratory experiment. Both are unashamedly decorative. The hazel is best grown as a single specimen; the white-stemmed rubus, whether planted individually or as a bold group, should be treated like a fruiting raspberry and the old wood cut down to allow new canes to take over.*

The white trunk and branches of Betula pubescens (B. alba) *appear luminous against a clear, blue winter sky. Even on a dull day, a single specimen or small group of these trees can be breathtaking, the silvery-white, light-reflecting surface of the bark giving off an almost unnatural gleam. To create a multi-stemmed effect, either plant three of these trees together in one hole or heavily prune a single specimen to within 15 cm (6 in) of ground level.*

Johnson shares my admiration: 'Birch bark is terribly important to me, both in winter and summer. The most relaxing gardening job I know is patiently peeling birch trunks. A young Kashmiri *B. utilis jacquemontii*, six years from seed, is already pipe-clay white like a Royal Marine's helmet.'

I am also very attached to my *Eucalyptus gunnii*. The devastating winter of 1983-4 appeared to have killed it, but fortunately it rejuvenated from the base, and lovely young foliage, silver-blue and rounded, was the reward.

The snow gum, *E. niphophila*, with its pythonlike skin, has an interesting trunk, but nothing like as striking as the cleaner white trunk of *E. pauciflora*. Looking out of the dining-room window of a friend's house one winter's day and seeing this tree was almost as exciting as sitting opposite a Monet or a Goya. It stood head and shoulders above its neighbours for beauty. In this small front garden – just far enough from the house, about 9m (30ft), for its whole form to be appreciated – it was particularly lovely.

Two gardening friends from South Carolina have written to me about the 'incredibly beautiful bark' of the crape myrtle, *Lagerstroemia indica*. They say: 'It sheds the outer skin in large exfoliates in the fall and by mid-winter the tree has a slick silver appearance. Since the majority of these trees are grown as multi-trunked specimens, they give the appearance of dancing maidens with their hair blowing in the winter wind. Daphne is supposed to have turned into a beautiful laurel, but we feel sure it must have been a crape myrtle.'

Among the broad-headed trees with interesting trunks, two acers are suitable for the small garden. *Acer grosseri* var. *hersii* has striking green bark striped with white, and *A. rufinerve* greyish bark striped with pink.

Pinus bungeana, the lace-bark pine, is remarkable in winter for its grey-brown bark. The bark exfoliates to reveal yellowy patches which shine out in the sunlight. As it ages, the trunk is said to turn almost white – this has certainly happened on very old trees in China, its native home. It usually has a multi-stemmed trunk and I can think of instances of it in several arboretums, especially the lovely specimen in the National Arboretum in Washington DC. With its light green needles, this pine contrasts well against broad-leaved evergreens such as camellias.

One of my best winter stems, and a must for any winter garden, is an ornamental bramble. The ghostly lines of *Rubus cockburnianus* are truly dramatic among other shrubs or in front of evergreens. Their strong yet arching stems make bigger and better clumps each year. It is the white bloom on its branches in winter that is so striking, not its leaves, nor its flowers later in the year, nor its fruit. Each spring you have to cut down the old stem to make way for new summer growth – just like autumn-fruiting raspberries. My original plants came from the Chelsea Physic Garden, where they had outgrown their allotted space.

Two of the willows also have an unusual and amazing white bloom on their stems in winter – do not finger them or you will take away some of their lustre. The vigorous shoots of *Salix daphnoides*, the violet willow, become veiled with a transparent white bloom over the deep purple bark; in the same way, the silver-grey hairs of its catkins conceal their purple hearts. This willow can be treated as a tree or pollarded into a shrub. Either way, the catkins are a wonderful sight in February against a blue sky or with the slanting sun illuminating them. Another of the salixes that covers its green stems with a white bloom is *S. irrorata*, which takes on an almost ethereal appearance, especially in fading winter light. It is most effective as a bush, achieved by pruning back in spring.

Foliage, variegated or uniform, is the principal provider of white and grey in the garden in winter. The white-edged leaf of one of my favourite shrubs, *Prunus lusitanica* 'Variegata' takes on a pretty pink flush in winter to match its ruby-red stems. It clips well, as does the silver form of privet, *Ligustrum sinense* 'Variegatum', but I prefer both untrimmed.

The leaves of *Rhamnus alaternus* 'Argenteovariegata', creamy and white-margined when you look at them closely, are grey from a distance. Used in a conventional way as a bush, this plant will lighten a mixed or shrub border all through the year. As a clipped ball or a standard it is perfect for tubs during the winter. I also like the white-margined *Ilex aquifolium* 'Silver Queen'; nearby, for contrast, you might plant *I.a.* 'Handsworth New Silver', with silver-margined leaves offset by purple stems.

A shrub with leaves that are creamy-white on green is *Griselinia littoralis* 'Dixon's Cream'. It is useful as a hedge, especially in seaside gardens. The tree purslane, *Atriplex halimus,* also an excellent seaside shrub, has one drawback – if the sparrows decide they enjoy its metallic, silvery-grey leaves, they will shred them. Its branches have a slightly lax habit, making an excellent addition to the mixed border.

The evergreen elaeagnus, *Elaeagnus macrophylla*, has shiny silver-grey leaves and fragrant silvery flowers early in winter. The flowers of *E. pungens* are also silvery, but it is the underside of its shiny green leaves which are dull white or grey. It looks well planted with the lovely *Picea pungens glauca* 'Koster', which has intensely silver-blue foliage. A small tree, it looks equally good at the back of a mixed border or as a single specimen on a lawn.

Silvery groundcoverers are a godsend in winter, particularly in darker spots where dense carpets of pure green might appear gloomy. Ivy, usually thought of as a green climber, has a number of variegated silver forms that make impeccable groundcover and also clipped edging. Top of my list comes *Hedera helix* 'Glacier', with its silver-grey leaves edged with a narrow margin of white. The smaller leaves of *H.h.* 'Adam' have an overall silver tinge.

Beware of the pale grey-green *Lamiastrum galeobdolon* 'Variegatum'. It can be too invasive, but if you select a place where it can run riot, you will enjoy its attractive netted leaves. If you shear the old stems back hard in autumn, its long arms will not send down roots; this will keep it under control and also make it flower better in spring. The three *Lamium maculatum* – 'Beacon Silver', 'White Nancy', and 'Shell Pink' – have all become deservedly popular. The first two have silvery leaves while the other is frosted green and white. They do well as groundcover all through the year and in winter the leaves develop deep pink markings when the temperature drops. We have planted *L.m.* 'Beacon Silver' under roses at Barnsley. After the lamium has flowered and begins to look tawdry, we shear it down to ground level and add some peat to give it a more cared-for appearance. After only a few days new leaves begin to sprout.

The patterned grey and green leaves of *Cyclamen coum* are also one of nature's joys, and with patience they will seed themselves to make lovely tapestried groundcover. *C. hederifolium* also has pretty mottled leaves, while those of *Arum italicum pictum* are marbled and veined with white. All three grow leaves in winter which disappear by summer.

For a bright silvery-grey carpet, I would plant *Alyssum saxatile* and *Cerastium tomentosum columnae*, which is fairly easily contained. Many of the completely grey-foliaged plants are low-growing. *Euphorbia myrsinites*, a groundcoverer with blue-grey leaves on prostrate stems, luxuriates in sunshine. A great gardener in Derbyshire, Jean Player, has planted it in her terrace garden, where it seeds itself prolifically, but I have failed to find the ideal spot for it in my garden – maybe I should try it in a stone trough, where it would have well-drained soil and could hang down over the sides.

These grey, spreading plants look particularly attractive with evergreen shrubs and border plants, or in partnership with red and purple foliage or golden leaves. The ones I choose for winter are an ivy with a grey tinge to its leaf, *Othonna cheirifolia*, even though its yellow summer flowers are not my favourites; the popular rabbits' ears, *Stachys byzantina* (syn. *S. lanata*); and a plant which my friend Peggy Munster liked, *Veronica cinerea*. This veronica is particularly suited to the edge of a border, falling over a gravel or stone path. Nor should you forget the grey leaves of dianthus; the one I favour is *Dianthus caryophyllus*.

Two biennials useful for their grey winter leaves are *Lychnis coronaria alba* – given the chance it will seed itself freely – and *Salvia argentea*, with large, hairy, velvety leaves. These low-level plants associate well with the slightly taller grey of *Senecio* 'Sunshine' and the silver santolinas, with grey sage, *Artemisia arborescens* and *A.* 'Powis Castle', and with the curry plant, *Helichrysum italicum* (syn. *H. angustifolium*) and the longer-leaved *H. plicatum*.

Several of the most attractive grasses have pale plumes or grey leaves, too. I always place them in a special category in my mind because the dense clusters of their leaves give them a solid appearance completely at variance with their feathery topknots. The well-known pampas grass, *Cortaderia selloana*, has silky white plumes over 3 m (10 ft) tall, carried in summer and retaining their freshness all winter long. I prefer it to the pink pampas, which looks unreal. For smaller plumes and leaves, I like *Pennisetum ulopecuroides*, the spikes of which, growing to 90 cm (36 in), have given it the name of fountain grass. Its leaves are grey and its plumes tawny yellow.

Miscanthus sinensis 'Gracillimus' has lovely feathery fruiting stems and leaves of a pretty grey-green. *Stipa gigantea*, the giant feather grass from Spain, with its grey foliage and huge flower heads that turn harvest yellow by autumn and last through the winter, is most impressive.

Red and Purple

Red and purple are the shock troops of winter: enjoy them individually; use them to heighten a group of whites and greys and golds; or, for a long-distance effect, plant them in a group – copper trunks, red berries, scarlet stems and rich purple foliage look startling together.

Left: Creating a fine sculptural shape, the fine white plumes and arching leaves of the pampas grass, Cortaderia selloana, *appear almost ghost-like in the mist and frost of an early winter morning.*

Below: The finely-textured, grey-green leaves of Senecio 'Sunshine' *are finely rimmed in white. On an overcast winter's day, a bush of this plant brings warmth and luminosity to the garden; under frost, the leaves glisten.*

Last winter, I was stopped in my tracks by a low hedge, or rather edging, of *Viburnum opulus* 'Nanum' in a Washington DC garden. Its dense mass of roseate twigs produced an overall crimson glow which brought immediate warmth to the cold look of the soil. And, here at Barnsley, sitting at my desk looking at the young red stems on top of my avenue of limes, *Tilia platyphyllos* 'Rubra', partly silhouetted against the heavy green of *Thuja plicata*, I see evidence enough that you do not need a whole army of red stems to make them stand out. What you do need is a green background to set them off. The brilliant red stems of *Cornus alba* 'Sibirica', placed in front of a dense evergreen, will take your breath away. *Alnus incana* 'Aurea' needs a green background, too, to show off its red-tinted catkins in January and its bright orange winter stems. And in the vegetable garden, the so-called red cabbages (really a rich, deep purple), planted next to moss-green, crinkly-leaved Savoy cabbages, will catch your eye each time you walk that way.

Once you start looking for shades of red in the garden in winter, you will begin to see them everywhere. Beautiful reddish-brown bark on the trunks and young branches of trees will start to seem an essential part of winter. Many of the acers qualify admirably for this role. *Acer davidii*, *A.d.* 'George Forrest' and *A. grosseri* var. *hersii* all came from China early this century, but much older in cultivation here in Britain is *A. pensylvanicum* from the east coast of North America. The variety *A.p.* 'Erythrocladum' was shown at the Royal Horticultural Society in London in the early spring of 1977. It was startling in its beauty. The young shoots are bright shrimp pink with a paler striation, and as a tree it is something to search out and plant. *A. palmatum* 'Sango Kaku' (syn. *A.p.* 'Senkaki' is another lovely small tree with a coral pink, stripy bark. One word of warning: these acers tend to be at their best for the first ten years, after which the trunks lose their striation and only the youngest, highest branches are effective.

I have recently discovered surprising reds among tree trunks. We tend to think of *Taxus baccata*, the English yew, as an evergreen which may be allowed to grow as it wishes or else be clipped into candle-flame or other, more exotic, shapes. We enjoy it for its dramatic emphasis or for providing a useful dark background. It was not until I read a passage on evergreens written by E. A. Bowles that I realized how exciting yew trunks could be made. On his visit to Glasnevin in Ireland one drizzling day, Bowles remarked, 'I was attracted to a row of glowing red-stemmed trees that showed up from a great distance. "Whatever are those trees?" I asked, "surely not *Arbutus andrachne*?" "They are yews," was the Director's reply. I thought they must be some scarlet-stemmed form of Irish yew then, but was assured it was nothing but a continued application of scrubbing brushes and elbow grease that produced those glowing trunks.'

So here at Barnsley we have tried to do the same. I am fortunate in having yew trees, which must have been planted in the middle of the last century, each side of the drive gates. The result will be a glowing burgundy red ... Come and see.

Among the shrubs with exciting rubicund stems, the willows come first. You do not have to find a damp place for willows, but if you can they will grow more effectively. The scarlet willow, *Salix alba* 'Britzensis' (syn. *S.a.* 'Chermesina') has conspicuous scarlet winter stems, and makes a grand stand along a river or stream. *Salix fargesii* is different – you might not guess it was a willow.

A more exotic touch of red is provided by the bright, hairy stems of *Rubus phoenicolasius*, the Japanese wineberry, which look even better in winter than in summer. Considering it is such a ready spreader, I am always surprised that so many of our garden visitors have not seen it before. The birds eat the berries and distribute the seeds around the garden, producing an abundance of new brambles coming up through other shrubs in the borders. They must, I think, sit on the iron railings nearby and leave their droppings on the stone paving – the wineberry comes up there in profusion.

Once the deciduous trees and shrubs have lost their leaves, the berries that crown them become all-important. Massed together, they briefly make cones of brilliant colour. By the time winter is fully fledged, many of them will have disappeared down the throats of the thrushes and blackbirds – the red ones seem to be the first to go. If I want berries on our holly for Christmas, I must net them before the tenth of December; if I am late, the birds will have got there before me. But the impact of those that remain is emphasized by their scarcity.

Among the mountain ashes, the red-berried specimens – *Sorbus* 'Embley' and *S. sargentiana* – lose their fruits first. The

One of the most arresting plants in my garden during winter months is the Japanese wineberry, Rubus phoenicolasius. *Frost highlights the hundreds of bristles on its red-brown stems; on sunny days the effect, although different, is equally exciting. The arching stems will grow to 1.8 or 2.5 m (6 to 8 ft) but need to be cut down each autumn to encourage new growth.*

Above: The translucent red berries of Viburnum opulus, *the guelder rose, persist well into winter and hang like pendulous rubies in the sharp winter light.*

Right: 'The fascination of the garden in winter is the magic of unexpected colour', writes Dick Balfour, former president of the Royal National Rose Society. Taken three days before Christmas, his photograph of Rosa *'Poppy Flash' – its petals and buds outlined in frost – has caught a moment of rare winter beauty.*

whitebeams – *S. aria, S. intermedia* and *S.* 'John Mitchell' – have larger, less conspicuous fruits, which the birds generally leave alone. I try to find a small basketful to lay out on a shallow dish as a table decoration in December.

Most of the fruits on the malus will have gone by December, but *Malus transitoria* has cherry-sized, red fruits which slowly lose their colour and hang on until the next autumn. Under the weight of its fruit this tree quite changes its contours, becoming almost weeping in outline.

We owe thanks to the Brooklyn Botanic Garden for giving us *M.* 'Red Jade', with its dramatic weeping branches and persistent red fruit. It is lovely all through the year – in spring and after leaf-fall with a multitude of red fruit on bare stems. It looks wonderful as a single specimen on a lawn, as does *Idesia polycarpa*, with bunches of red berries lasting throughout winter. It is important to remember that trees of both sexes are needed for fruiting.

The broad-leaved evergreen shrubs yield a rich crop of berries too. *Viburnum rhytidophyllum* is a wonderful sound barrier and a fast grower – after a few years it will reach 2.5 to 3m (8 to 10ft); if you have both sexes, you can expect a winter display of red berries which gradually turn black.

The red or yellow berries of pyracantha are usually left by the birds until they have first devoured the holly and sorbus fruits. Is this because they do not ripen so soon? Pyracantha, incidentally, makes just as good a bush as a wall shrub. *Cotoneaster* 'Cornubia' and *C.* 'Hybridus Pendulus' also keep their sparkling red berries, and the weeping form is excellent for an important position, or used in pairs to flank a gateway or mark the top of steps.

Stranvaesia in all its forms is distinctive for its matt-surfaced scarlet berries. *Stranvaesia davidiana* (now called *Photinia davidiana*), which grows emphatically upright, is a useful shrub for all seasons, but as autumn turns to winter, the leathery green leaves are enriched with the reds and yellows of its ripening berries. Their vigorous growth can be kept to the height you require by hard or annual pruning. The birds seem to ignore their berries, and last year paid no attention at all to those on the yellow form which we had fruiting.

When skimmias are good they are very very good, but you need persistence in acquiring good berrying forms. They are slow-growing and take time to establish, but their brilliant red berries make perseverance well worthwhile. Hilliers Nurseries of Winchester in Hampshire recommends *Skimmia japonica* 'Nymans', a female which has an eventual height of

1.8m (6ft) and berries freely. If you choose this specimen, you should remember to plant a male among the females to encourage berrying.

Do not immediately dismiss laurels from your mind because they had the misfortune to be overplanted in Victorian gardens. If you are keen on winter berries, then the variety *Aucuba japonica* 'Salicifolia' has striking red fruits, and is resistant to town pollution – if you live in an urban area with a large garden, they can give good service. My advice is to choose one with a bold leaf.

For those who garden in acid soil, the deciduous American winterberry, *Ilex verticillata* 'Winter Red', should rank among the top shrubs for winter berries. They look especially good as a group near the house, if space will allow. Their abundance of small, sparkling red berries is even more dramatic than those on the evergreen varieties. It was selected by Bob Simpson of Indiana in 1960 from a batch of seedlings growing in a nursery in Pittsville. Utterly hardy, it will also thrive as far south as Florida. I have seen it growing in Delaware, looking spectacular in association with red-stemmed cornus, and covered in fruit as late as March. Also deciduous and with berries like small purple jewels which are much more unusual, is *Callicarpa bodinieri giraldii*. Once established, it is a great addition to the winter garden. The best specimens I have seen are in the Brooklyn Botanic Garden in New York.

Those other targets for the birds, hips – or 'heps' as they are sometimes called – are not as luxuriant in their fruiting as berries. In the hedgerow there is something grudging in the way they push out their knobbly red fruits at the ends of bony stems. The birds are canny about their offerings – have you noticed how they will go for the succulent hips first and will only try for the tougher, less juicy ones when the winter is exceptionally harsh?

In the garden, roses have hips to offer, often low-key, sometimes prolific. Dick Balfour tells me that in his Essex garden the longest lasting – and he grows a great many roses – are those of the repeat-flowering climbers, such as 'Dortmund', 'Pink Perpetue', 'Aloha' and 'Morning Jewel'. Their hips are all upright, but when you look at many of the shrub roses you will find theirs hang down on bending stems. I enjoy watching the birds feasting on the profusion of juicy hips on my Rugosa hedge. Its stems are flexible and its hips pendant, so it requires an acrobatic performance from the birds. By mid-December, they will have cleared the lot. Some years the Rambler rose 'Wickwar' will keep its profusion of

small hips right through the winter in my garden – they seem to be too hard for the birds to enjoy.

Red leaves are softer in outline and more extensive in area than red berries. Some shrubs have both. I have already mentioned *Stranvaesia davidiana* for its distinctive, matt berries, but when we were making the red end to our border we used it for its shiny evergreen leaves, the oldest of which turn a fiery red in winter and usually hang on until spring. This useful shrub grows tall and has a distinctly upright habit. It only takes up a couple of metres (yards) of space and is fairly see-through, because the distance between each leaf cluster is long. Do not be nervous of clipping its longest branches back, but do this in early spring; by June new growth will be on its way and the cuts will not show.

The winter foliage of some plants retains its red and purple shades, and many of the evergreens – ivies, bergenias and cryptomerias – become bronzed. I was sad to lose the winter bronze effect of my *Cryptomeria japonica*, which we had to take out when it outgrew its allotted space. Pippa Rakusen has written of this tree, the Japanese cedar, that it is 'particularly striking when the winter foliage is quite purple-red'.

The red chokeberry, *Aronia arbutifolia*, is a medium-sized shrub with brilliant red leaves which I have seen growing in the United States. Its deciduous leaves are spectacular in autumn and last well into winter. A real blood-red that I must try to include in my borders I have also seen only in America. It is *Imperata cylindrica rubra* 'Red Baron', the Japanese blood grass, a perennial which looks striking set against grey or green neighbours. The whole colour range – from deep red to bronze-purple – is to be seen on the rather leathery leaves of *Leucothoe fontanesiana,* a small shrub with a preference for acid soil.

Several groundcoverers have red and purple in their foliage, too. The leaves of the epimediums, often marbled with reds and purples, are the plant's greatest asset. From autumn through winter, they have soft tones of bronze-red and coral. Harold Epstein, past President of the American Alpine Garden Society, has many varieties growing in his garden in New York; seeing them in earliest spring, I was amazed at the diversity of forms he has collected. This is the excitement of gardening – always another field to discover, to specialize in, to enjoy. Epimediums are basically woodland plants – they will survive in poor, dry soil, but give them some rich humus or leaf mould and they will respond as exciting groundcover.

The tiarellas love shade and a rich soil, too. My favourite is *Tiarella wherryi* for its pinky-green leaves like soft velvet. They look superb in the front of the border if you can arrange enough shade for them. *Tellima grandiflora rubra* is at first sight less spectacular but has many virtues: it is not invasive, the flowers are pretty in summer, and, most important, the leaves become beautiful in winter after the frost has tinged and coloured them.

Twenty years ago Nancy Lindsay supplied me with *Euphorbia amygdaloides* 'Superba', an especially good form of the native wood spurge. It has done well for us but hates the cold east winds of March, which sometimes spoil its promising flower spikes. A better form is *E.a. rubra*, with red stems and purplish leaves.

For reliability and handsome winter foliage, I class the bergenias with the euphorbias. But if you are wanting red leaves in winter, you should choose *Bergenia* 'Abendglut', *B. x schmidtii*, 'Sunningdale' or *B. cordifolia purpurea*.

The leaves of *Viola riviniana* are a lovely dark purple and it carries a fine quantity of pale purple-blue flowers. I was given one by a friend with a small Oxford garden who told me that it would spread everywhere. We put it in our border, where it made a reluctant increase – I was always hoping for more. Then somehow it arrived in the vegetable garden, probably carried in on its own compost, and during the past two summers has just rampaged. I am not grumbling; I enjoy encouraging it and selling the overflow for others to enjoy.

The purple-leaved clover, *Trifolium repens* 'Purpurascens Quadrifolium', is another low plant which also sells well in our nursery. It has the added psychological attraction of frequently having four leaves. Why this should be thought so lucky I am not sure, but I do know that it makes wonderful groundcover. It is devoured by uninvited hares and is one of the few plants I send by post – to ladies or gentlemen wishing to give good fortune to their friends. To me it is just the plant to round off a corner of my herbaceous bed and allow early irises to grow through – those fresh pale purples look lovely framed by clover leaves. The luck lies in the eyes of the beholder – you and me.

Another low border plant with red-to-purple leaves is the sedum. Gardeners either love all things small or simply cannot find the right place, however much they admire them. I fall headlong into the last category. Perhaps in my late seventies my horizons will change. Why not? Life evolves and so do plants, and I am already leaning towards *Sedum spurium* 'Atropurpureum'. Beth Chatto says, 'The foliage is so alluring it hardly needs the rose-red flowers.'

A plant with bronze-red leaves of which I am particularly fond is *Heuchera micrantha diversifolia* 'Palace Purple'. A

Purple and blue are to be found in winter among the hellebores, the scillas and the reticulata irises; pink, in spite of the cyclamens and other notable exceptions, is a rarer accent at this time of year. In Princess Sturdza's garden at Le Vasterival near Dieppe, the varied shades of pink of Hydrangea serrata *'Preziosa',* Phormium tenax *and the panicled seed heads of the sedum are both bright and delicate.*

winner, its beautiful heart-shaped leaves cluster round each other. You can increase it by seed and divison, but save only the most richly coloured of the offspring. A single plant is eye-catching, and a group can be spectacular.

If you aim for movement as well as colour in your garden, you must have grasses. The purple moor grass, *Molinia caerulea* 'Variegata', makes shapely tufts, 60 cm (24 in) high, with flower spikes which keep their freshness well into winter. *Miscanthus sinensis purpurascens*, as its name implies, has leaves turning to a rich purple in autumn, topped by reddish plumes. Caught in sun when the wind is blowing, or rising from a bed of snow, they show that winter has a lighter, less serious touch.

Black

Black and darkest grey are the colours that first come to mind for most people when they think of winter. Their thoughts turn to lowering clouds, long dark nights and the stark outline of trees and bushes. All they remember are the dreary expanses of sombre shades that they see from their windows.

For the gardener, however, black is a marvellous background colour, an essential factor in creating dramatic effects. I love the silhouettes of trees, the emphatic nature of their bare tracery and the strong shadows cast by their trunks. As a contrast, think of the flat, shiny surface of water, which in winter looks like polished slate. There is no finer winter sight than trees reflected in water, black on black.

You can group trees where they will have the most impact

Left: Snow transforms the swimming pool garden at Westwell Manor in Oxfordshire. With the reflections of the twin hexagonal huts clearly visible on the water surface and the scalloped edges of the pool emphasized by a fall of snow, the picture catches a magic moment, quiet and still. It is a study in black and white, with only the golden globes on the changing houses providing colour.

Right: On a snowy day a small back yard in Ham, London, takes on the distinctive qualities of a black and white photograph. The varying shades and hues of flowers, berries, leaves and grass are all disguised, leaving only the stark silhouettes.

and plant colourful shrubs and groundcover perennials within the framework they provide. Climbers can use trees as a natural trellis; shrubs can follow the contours of a dark pond and be reflected in its water.

Nature herself has recognized the supporting role of black: there are very few plants which are black through and through. I can think of one, though, which I have planted in my garden – *Ophiopogon planiscapus nigrescens*. It looks best surrounded by a sprinkling of pale gravel or chippings to show off the black leaves.

Then there are the less obvious blacks – the stems, berries and catkins. *Cornus alba* 'Kesselringii' has deep purple stems which look black in certain lights. Chestnuts lying on the ground, so richly mahogany-coloured in autumn, turn black at the onset of winter. *Ilex crenata* has black berries and those on the arborescent ivies become black as winter moves on. For amusing and unusual black catkins towards the end of winter, try planting *Salix melanostachys* – and with a stretch of imagination, the very darkest flowers of *Helleborus orientalis* can appear black in the evening light.

WORK & PLEASURE

WINTER JOBS never seem as urgent as summer ones. They can always wait until the spirit moves, the weather is right, and a free day comes along. But this is a dangerous fallacy. First comes Christmas, then a frost, then a fall of snow, then a holiday... suddenly spring is round the corner and the jobs are still waiting to be done. From past experience I know that if I want an exciting and worthwhile garden through all four seasons, I must not neglect my winter work. I must make sure that routine jobs are completed and extra projects tackled.

I like to categorize the jobs according to the weather in which they can be done. Then you can turn to your lists and be sure to find a job that suits both you and the weather on that particular day.

Many of the categories overlap – you can do your wall pruning and tying in when the ground is too wet to work or when the snow is lying. So please do not try to put your work programme onto a computer. Nature cannot be regulated – 'By the second week in December all the double digging must be finished.' Things just do not happen like that; we learn to trust our basic knowledge, our instinct and the feel of the soil.

For example, there is no substitute for going out into the garden on windy days if you want to find which places take the full brunt of the blasts and which are relatively protected. Now is the time to plan a shelter belt. It can be effective surprisingly quickly, especially if you plant fast-growing conifers in front of or among slower-growing hardwoods. Later on, the conifers can be cut when they become too tall and oppressive, allowing the hardwoods to come to the fore.

The key to winter work is choosing the right day. Then some so-called jobs are more of a pleasure than a chore. Make the most of those moments of happiness when one is not so much doing a job as enjoying a particular sensation.

I love walking in my greenhouses on Sunday mornings in mid-winter. I know I will be alone and can slowly pinch each leaf on the scented geraniums before going to find out what is happening in the mist propagating house. I will put half-canes into the trays of cuttings which are ready to be taken off the spray bench on Monday morning, and note which cuttings have been impossibly slow to make roots. What have we done wrong? There is time to ponder.

Valerie Finnis, a great gardener and specialist in alpines, loves opening her cuttings frame, kept warm under a thick blanket of snow, after a thaw and finding everything inside looking well, ready to be potted to make another generation of plants for her own garden or those of her visitors.

The American gardener George Abraham knew much the same kind of pleasures; 'There's something soothing about firming seeds in the soil and tending plants under glass while raindrops and snowflakes fall outside against the panes.' I am sure he did not think of these jobs as work – they were all part of his pleasure.

Building a bonfire for me is another pleasurable sensation. Once started, it must be kept smouldering for days on end – at any rate on those dry sunny days when the wind is blowing gently, just enough to fan the flames. You must tend it last thing in the evening. Give it a good feeding, so that when you go out again the next morning to see how it is faring, you will be able to push the sides into the still-hot centre. The dry wood will crackle and the flames spring into life once more. Only put bad weeds on the bonfire; the others must go on the compost heap so that their nutrients will eventually be returned to the soil. Bonfire smoke, with its aromatic and atavistic smell, will drift over your garden. Almost like a drug, it will halt you in your tracks, and remind you of your childhood. A funny thing is that other people's bonfire smoke is always perfectly awful when it drifts across your garden.

Walking round the garden on a brilliantly sunny winter

A garden covered with snow is often enough to bring even the keenest gardeners inside, especially if there are plants to tend and propagate in the warmth of a heated greenhouse or conservatory. In winter, indoor plants need as much light as possible so expanses of glass should be kept clean to allow in every ray of sunshine. This plant-filled glasshouse makes an ideal place for bird-watching.

morning is not work, but it is time well spent if it gives you ideas for special projects and allows you to notice and change things which are not quite right. Some of my happiest moments are spent standing in the garden, seeing where improvements could be made and new ideas carried out.

On those truly awful days when there is nothing for it but to stay inside in front of a warm fire, I like to read old classic gardening books. Life styles have altered but the vagaries of the winter weather take us as much by surprise as they did gardeners of the past.

Gilbert White kept his diary from 1751 for twenty years. He recorded in his *Garden Kalendar* the work carried out at Selborne in Hampshire and also entered, day by day, the temperature, the barometric pressure, the direction of the wind (often at 8 a.m., noon, 4 p.m. and 8 p.m.), the rainfall and a summary of the weather – for instance, 'great dew, bright, showers about, sultry'. With equal fervour he observed when trees first came into leaf, when plants first flowered and when fungi first appeared. His observation of nature was minutely detailed, whether it was of birds – their movement and their appearance – or of insects, of fish or of other animals.

Philip Miller, 'Gardener to the Worshipful Company of Apothecaries at their Botanic Garden at Chelsea', first published his much-read volume in 1732; the seventeenth, and final, edition came out in 1792. It is a Kalendar of the works necessary to be done in the kitchen, fruit and flower gardens. December, he writes, is the darkest month of the whole year, and 'is subject to different sorts of weather: sometimes the ground is frozen up, so that little can be done in the garden: and at other times there are hard rains and thick fogs, which render it very uncomfortable stirring abroad, but especially to persons of tender constitutions: and this weather is very injurious to tender plants'. At this time of year, he recommends you pick snails out of the holes of old walls or from under pales, hedges, broken pots or other rubbish.

Across the Atlantic, Thomas Jefferson, third President of the United States, and a countryman at heart, started his *Garden Book* in 1766 and continued until 1824. He loved his garden and estate, so it teaches us much about his gardening activities and the plants he introduced into his country, many of which came from Europe.

I have recently come across two chronicles which are remarkable because they provide immensely detailed pictures of, in one case, a countryman's eager anticipation of spring, and in the other a truly creative gardener's plan of work and catalogue of achievement. The countryman, Mr Masham, lived and worked in 18th-century Gloucestershire and for sixty-two years recorded faithfully all the variations in the events that herald the arrival of spring. Henry Hucks Gibbs, later Lord Aldenham, kept a garden year book from 1869 for thirty-three years, in which he recorded the trials and triumphs in the redesign and replanting of his garden and park at Aldenham, Hertfordshire.

These old records are never a waste of time. Tools, chemicals and the plants themselves may have changed and developed over the years, but the soil remains the same and much of the advice they give remains relevant. The gardeners of the past recognized that, whatever the weather, there is plenty of real work to be done.

Foul Weather Jobs

Foul-weather work starts with husbandry – drawing up a maintenance programme and a shopping list of equipment. Flower pots must be washed and new ones ordered. Tools have to be sharpened, cleaned and given new handles. Machinery needs servicing, electric cables should be tested and the hose pipes checked. We often find we are short of joiners.

Include in your shopping list any twine, bamboos, wire ties or nails which may be in low supply, grit or gravel to spread over slippery places. Now is the moment, too, to make sure you have enough spray for treating the lawn, so that you are prepared for that morning in early spring when the weather is perfect for applying weed-killer and fertilizer. Also, calculate what fertilizers and sprays you need for your borders and vegetable garden, and make sure you have liquid feed for your pot plants.

On a wet day in early winter I like to consider what paint we will need to smarten up doors, frame tops, gates and railings, so that when a suitable wet or frosty day comes everything will be ready and we will not have to waste time making a special journey to fetch supplies. Then there are the paint brushes, abrasive paper for rubbing down and brush cleaner for the end of the day. To make sure that the seats and garden chairs which need repainting will be dry on the day, we put them under cover early on.

Make sure that your greenhouse is getting as much daylight as possible. White walls reflect light. You will probably have attended to them in the summer; if not, white over any areas you can now. I am always surprised how much lighter it becomes inside after the glass has had a good clean. In the greenhouse there is always work to be done – potting on when essential, restaking and tying in any climbers which have taken off, keeping all the plants happy with the right temperature, food and moisture. Spraying too is important. However meticulously tidy and clean your greenhouse, white fly, greenfly and scale insects seem to enter. Prevention is much easier than cure. Our routine is quite simple and has proved its worth over the years. Every Monday, or failing that, Tuesday, plants are sprayed with a proprietory brand insecticide. We have three different brands, including systemic. These we use in rotation to prevent a build-up of immunity. You must be regular because of the life cycle of the insects. A word of advice – some plants may be allergic to some sprays (you will soon discover which). These plants must either be removed temporarily or covered over with newspapers. Young seedlings should also be covered.

Prepare for seed sowing by making sure you have enough small, inexpensive envelopes for seed sorting, enough seed trays, individual plantpaks and seed compost. I like to use fresh seed compost, so I mix in any left-over from last year with potting compost. You will need a packet of seed dressing, too, to discourage seedlings from damping off. Then, on the wettest or coldest day you can embark on one of the most rewarding gardening jobs – sorting the seeds.

We collect seeds during the year from gardens abroad as well as from our own. Our routine of seed collecting is haphazard and will shock the purist. We snip the seed heads off when the seeds are ripe and put them straight into labelled paper bags or plastic cream pots. As this happens at the busiest time of year, they have to stay in the dry atmosphere of the kitchen until a wet or frozen winter's day. Then we cover the kitchen table with newspaper and the fun begins. Like Penny Hobhouse, we sort and clean them 'in gloating anticipation'.

With the small seeds, a fine sieve helps to separate the chaff, but we do not clean them to the seed merchants' standard of perfection. Sometimes good finger nails are needed to crush the capsules and release the seeds, but generally they will come tumbling out. As with all gardening, time is important, so devise an easy way to tip the seeds into the envelopes. Label, date and seal them. Separate the flowers from the vegetables and stack the envelopes in shoe boxes, making sure that they are in alphabetical order. Then list the seeds and note beside

Tidying the garden shed is a job for a rainy day. As well as bringing order to chaos, it will remind you to buy nails, labels, twine – some of the many things you will need for the coming months – and to oil and mend garden tools and machinery so everything is in good order when you need it.

each how and when they should be sown: in the greenhouse, in drills in the vegetable garden, directly into the beds.

After you have cleaned and sorted the seeds, there is hardly a moment before it is time to sow them. Looking through my diary, I find that we have always sown the first seeds – lettuce, cauliflower and parsley – late in January. Two seeds go into each individual compartment of a plantpak, so that when they are potted on the roots are not disturbed.

Serious seed sowing starts in February, usually in trays, sometimes individually, a few each day until the space where they are to germinate is full. Do not forget the seed dressing. Each tray is then labelled and dated, put in a clear polythene bag and sealed to retain the moisture. The trays are taken down to the cellar to be kept warm and at a reasonably constant temperature by the boiler. Some seeds require light to germinate; others are covered with newspaper to keep them in the dark and, as they germinate, are removed from their bags and placed under 'daylight' lamps in another part of the cellar. There they will stay and grow on until they are ready to be pricked out. Seedlings should never be allowed to suffer a setback because of drying out or a sudden drop in temperature.

We have kept careful records of all the seeds sown this way indoors since January 1979. The first column shows the date sown, the second the date germinated. A third is reserved for comments – own seed, bought seed, seed from other gardens. The encouragement comes when germination takes place within five days – the usual time for cabbage and cauliflowers, sweet peas and sunflowers. When germination takes longer than it should, we try sowing a new batch and leaving them in the light. This happens regularly with *Nicotiana sylvestris*. Our own findings are invaluable, both as a factual record and as an *aide-memoire* to ensure that none of the plants we need for the coming seasons has been forgotten.

There are sure to be other foul-weather jobs, such as rearranging sheds and inventing ingenious ways to store bamboos and pots – and every garden will have different ones.

Icy Weather Jobs

Icy-weather jobs are part defensive – protecting the garden against the weather – and part creative – genuine and enjoyable work. Once you have taken care of the essential tasks – clearing paths, for example, and checking the lagging on outside water taps – you can enjoy the beauty that ice and snow bring to your garden. You may want to get out your camera. But even if below-zero temperatures go on for weeks on end, there are still some real gardening jobs to be done.

Charlie Gale, who runs his nursery in Pennsylvania with his son Chuck, catches up with some of his pruning jobs. In England we are made to feel guilty if we do our pruning when the sap in the branches is frozen, but Charlie prunes all winter long, whenever he has time. 'I prune even when there are icicles on my own ears and eyebrows,' he says. 'I am not going to harm anything.' He 'opens up' his trees and shrubs, then stands back and admires them. He says he loves the sculpted look they take on when the snow lies on the deciduous branches and on the re-berrying plants – *Ilex verticillata, Viburnum dilatatum, V. opulus* and aronia.

We first prune our espaliered and shaped apple trees in mid-August; any further tidying-up that needs to be done is a winter job. The old standard apples and pears are pruned in January or February, not wasting a day when we could get to the borders or vegetable garden. I love the shaping and pruning of the trees in the wilderness part of my garden – the sorbus and prunus, amelanchier and poplar.

Pruning is not a job to be hurried; you must ponder before taking off each branch. Remember that you can always cut off more but you cannot put it back. Take away any crossing branches, see that the trees are well-balanced and keep the trunks clear of side shoots which hide the beauty of an attractive trunk. Pollard or coppice the willows and keep the variegated-leaved *Populus candicans* 'Aurora' to a mere 3 to 3.6m (10 to 12ft) so that the young leaves with their pretty markings are well in view in summer. I like to be able to see through the branches of the nut trees, so all last year's new growth is cut off. Some winters, we do not get round to this as well as we should and then it annoys me all summer. 'It is worth any amount of effort to be able to see your house through the arch of a tree,' wrote Thomas D. Church.

Beautiful and consistently hardy, a bold specimen or hedge of Ilex opaca *is an impressive sight with its rich green glossy leaves and abundance of bright red berries. Native to the United States, the opaca holly is particularly striking for its richness of colour. It is available to gardeners in Britain. One of the broad-leaved evergreen shrubs that can be pruned back in all weathers, holly is an essential ingredient of traditional Christmas decorations.*

By rights summer-flowering shrubs – weigelas, philadelphus, kolkwitzias, spiraeas – should be pruned soon after they have flowered. This keeps them in good shape and allows time for new shoots to grow before winter. If you did not do this, it is wise to take out some of the old wood now to encourage growth in the spring.

Potentillas, deciduous cotoneasters, Japanese maples and hamamelis can be shaped and any old wood cut out in February. I like to use *Potentilla fruticosa* in varying shades of yellow in the front of my borders, and they may need to be reduced in size considerably every three years; February is the time to do this, too. I enjoy the brown, rather furry look of the dead flowerheads in winter, so I leave them as late as possible.

Fair-weather Jobs

Wall pruning is a job for which you will need a warm coat and a sunny day. To make the best possible use of walls, you will have quite a lot of pruning, tidying and tying in to do in winter.

Our longest wall faces north-west, not the most favourable for many sun-loving shrubs and climbers, but it is interesting most of the year. Very little sun reaches it until midday in winter, so it is quite cold for the fingers working there in the morning. We tend to choose a sunny afternoon.

You will need a wheelbarrow and a sheet, one for bonfire material and the other for composting. An apron or a coat with poacher's pockets is an advantage, as you will be changing tools constantly. You will also need secateurs and long-handled pruners, a sharp knife, nails (preferably non-ferrous) and a hammer, and tarred twine as well as 'twist ties'. Solid steps are essential, and if the soil is wet you may need boards to put under the feet of the steps to stabilize them.

On our north-west wall we have climbing roses, clematis and wall shrubs such as itea, winter jasmine, honeysuckles and ivies, but in addition we have other shrubs – deutzias, *Ribes speciosum*, buddlejas – normally grown free-standing,

Acer japonicum, a Japanese maple, does not require drastic pruning. It naturally forms an attractive shape, although a few branches cut for the house may give your bush an added elegance. As with any tree or shrub, stand back, pause a while and consider well before you start pruning. Remember that snow-laden branches will hang and change the shape of your tree so shake them free.

which appreciate the protection of a wall and can be trained, but not too vigorously, against it. These are not winter-flowering plants, but most need attention during the winter.

The roses have their routine pruning, the honeysuckles and ivies are kept in control, the *Ribes speciosum* and ceanothus are fanned out neatly. We usually prune the buddlejas back to halfway in November, just to make them look tidy, then we give them a severe pruning in March. The clematis all need different treatments, and we find it a great help if we write 'hard', 'light' or 'none' on the back of the labels to remind us what pruning they require in winter.

Wall pruning may not seem so crucial as shaping trees but it is a job on which it is worth spending plenty of time and taking great care – remember that a false cut in winter is a friend lost in summer. Pruning also has its pleasures – at the beginning of the day, standing back and assessing what needs to be done and then, when all is over, standing back again and admiring the neatly trained shapes you have achieved.

There are many wall shrubs and climbers that can be trained on a wall and you will simply have to make your own choice, and get to know the best treatment for each. *Cotoneaster horizontalis* is usually grown as groundcover, but its branches grow in such an attractive arching way that, pruned and tied back against a wall they can make a good vertical pattern. *Abutilon vitifolium* benefits from wall protection; you should tie back some of its branches and leave others to billow forward. *Hydrangea petiolaris* will need only a general tidy, snipping off last year's flower heads, but *Parthenocissus henryana* and *Polygonum baldschuanicum* will both take over and become much too rampant unless you are firm with them.

Bed and border clearance depends on the climate as well as the weather. In Britain we should have begun to clear the borders by December 1st, but in the colder states of America the work must be started much sooner and finished before Thanksgiving Day.

In an area where the temperature oscillates from well below freezing to 40° or 50°F (4 – 10°C), it is important to prevent as much 'heaving' of your herbaceous plants as possible. Shallow-rooted plants suffer by being lifted partially out of the ground, becoming loose and exposing some of their roots. Wait until after the first frost, then lay a good mulch of salt hay or a generous covering of evergreen branches on the frozen ground, trapping the frost into the ground and keeping the plants firm. The mulch helps to protect any leaves above ground from the severest frost and the winds.

In a less extreme climate it is unnecessary to cut down anything that is still looking green and attractive. But a start must be made before the first hard frost. Ideally, you should work systematically, from one end of a border to the other. In my experience it does not work out quite like that. One of the reasons is the bulbs. Our method is to have areas in the beds where patches of tulips have been planted to be followed by annuals or half-hardy perennials: forget-me-nots, wallflowers or polyanthus. To get the spacing right we put the plants in first and then interplant with the bulbs, about nine to the square metre or yard. We give the whole area a top dressing of leaf mould, mushroom compost or peat mixed with a small amount of slow-release fertilizer. This will remind us that here are bulbs, as yet unseen; it is so easy to forget exactly where we planted them.

As I try to have groups of tulips and other bulbs all round the garden, we have to move to the next bed and plant these before doing the final tidying up. I like putting 'ribbons' of large-flowered Dutch crocuses in quantity down the centres of the beds, and the delicate *Iris histrioides* and *I. reticulata* at the corners. A mixture of the free-flowering species *Crocus chrysanthus* will push through a groundcover of *Ajuga reptans* 'Multicolor'. They need to be near the edge of the borders so that you can enjoy their delicate beauty.

Then there are the primulas and cowslips. We try to dig them up after they have flowered in the spring, divide them and line them out in the vegetable garden. We have to remember to put them back at the front of the bed at the beginning of winter – the primulas, especially the Barnhaven strains, will often send up a few winter flowers. The tougher grape hyacinths and scillas are left each year and come up through low saxifrages and around the base of deciduous shrubs.

Before you start your final work on the borders, think about each one carefully. Which plants need dividing? Where can improvements be made? This is the moment to remind yourself of the spring and summer look of the beds by turning to any photographs you might have taken. Try to achieve a succession of colour through the seasons.

Patterns are often more important in winter than they are in summer. These box balls lining a brick path at Barnsley add interest to a border when herbaceous plants have been cut low; in summer they are almost submerged in foliage. Rosa 'Veilchenblau', Akebia quinata and Lonicera 'Dropmore Scarlet' have been pruned back against the wall.

Above: The seed heads of the grey-leaved Phlomis fruticosa *catch the frost and attract the birds. Left unpruned, this shrub tends to sprawl but you can cut it hard back to the new buds, even down to one third. Do this in late spring and it will soon be covered with leaves.*

Right: In the Royal Botanic Gardens at Kew, clipped hornbeams form an impressive hedge on stilts, underplanted with snowdrops for winter interest and cyclamen for late summer colour. Regular clipping ensures that the raised hedge retains its shape.

Having decided which plants need to be moved, choose a day when the soil is right – it will make the work much easier and more enjoyable. Plants hate to be shifted in damp or soggy ground – if the earth sticks to your boots it is a sure sign that it is too heavy to work and that wherever you tread you will pan down the soil – quite the worst thing you could do. If for any reason you have to work on wet soil, you should stand on a board to disperse your weight.

Before you start, get all your equipment ready: an empty wheelbarrow for weeds and spent stalks, a spade, a fork, a trowel, secateurs, short bamboos in case you want to mark a special plant, a few labels and a pencil. An extra bucket is always handy – it helps you to keep the lawn or path clean if you lay down sacking or sheets of polythene where you are working, and we always put boards on the grass edge to protect it.

Work through systematically, cutting some of the herbaceous stems low and leaving those which keep attractive seed heads through the winter. We cut the other stems very low, but I can appreciate why in colder climates it is better to leave 10 to 12cm (4 to 5in) of stub, for these will help to anchor the winter mulch of salt hay, leaves, pine needles and evergreen boughs. Lift out the plants you have planned to divide, and separate them using two forks back to back. Keep the best pieces and discard the rest or put them on one side to give away or use elsewhere. Some biennials will have seeded themselves – evening primroses, mulleins, foxgloves, honesty. Look out for these and save them to fill in any gaps.

The plants we save for their attractive winter seed heads are acanthus, bergamot, sedums and the herbaceous *Phlomis russeliana*. John Metcalf of Four Seasons Nursery in Norfolk has kindly given me some more suggestions. They include *Chelone obliqua*, agapanthus, *Eupatorium purpureum*, ferula, *Galtonia candicans*, helianthus and several of the grasses.

James van Sweden has original thoughts on planting: 'I choose plant species that have value throughout the year in leaf, flower, seeding habit and overall form. I set them out in masses. Instead of trimming off all of the spent flowers and dried leaves, they are left in place through the winter months. The winter look features the drama of the dried grass stems with long seed spikes, the dark heads of Black-eyed Susans and the bronze flowers of *Sedum* 'Autumn Joy'. I want my winter gardens to look like a big dried bouquet full of flowers, leaves and grasses.'

You may not share this philosophy and may prefer to clear your beds. As each bed is finally tidied, you should give it a

February – scratching about making emerging bulbs and hellebores look nice above clean earth or compost.' On a February day, it is rewarding to go round and very carefully cut off and remove any old, browning, untidy leaves from your hellebores. This will reveal the flowers as they open and add a touch of perfection.

Work in the Vegetable Garden

The vegetable garden and the flower garden should not be in competition; we should enjoy both equally. William Lawson, the early 17th-century clergyman who delighted in all the pleasures of his garden, wrote in *The Country House-wife's Garden:* 'Herbs are of two sorts, and therefore it is meet that we have two gardens: a garden for flowers and a Kitchen garden ... Your garden flowers shall suffer some disgrace, if among them you intermingle onions, parsnips, etc ...'

Here at Barnsley we have designed our vegetable garden to give us as much aesthetic pleasure as the flower garden. It has an overall framework of paths for patterns and fruit trees for height – the essential dimension. The French sometimes have borders, 90 cm (36 in) wide each side of the main paths leading through their potagers. These they fill with annuals and perennials for cutting. The permanent framework is supplied by espaliered fruit trees planted as backing to these borders. Our potager does not include cutting borders, chiefly because of lack of space. Permanent decorative features as you enter the garden are created by twin lavender hedges. Recently, I have planted ten standard gooseberries as intermediate supporters, standard roses form a circle in the centre, and trelliswork arbours are on two sides.

Early each winter, we stand in the potager and make notes of where next season's vegetables will be sown and grown. I like to do this consulting the *genius loci* – I can imagine more easily what the overall effect will be. First we have to decide where the sweet-pea arches will go, and manure their earth for good results. Then we have to decide where to put the climbing beans, especially the scarlet runners – pretty enough in blossom to be in the flower garden – and the 'Blue Lake' beans with their bicoloured flowers, which always draw comment from visitors and are good for eating, too. We have a winter work programme (shown overleaf) to which we turn for guidance, but it has to be flexible as the frosts and snows come and go.

further top dressing or mulch of leaf mould, mushroom compost, manure or peat with fertilizer. This feeding of the soil is important: you cannot expect the best results and displays from your plants on a starvation diet. After you have cleared away your tools, sheets and wheelbarrow, get a rake or brush and clean up your turf, stone path or whatever edges your border. This is the final touch.

As the winter unfolds and the bulbs begin to push through, it is especially satisfying after the snow has melted to spend a few moments pricking over the soil. Penny Hobhouse wrote to me: 'I love the winter tidy of perennials and then again in

A Winter Work Programme

December

Dig and manure areas as they become empty.

Take all the yellowing leaves off the various brassicas: they can harbour bugs and encourage disease. They also look unattractive.

Cover the globe artichokes with a layer of straw and perforated polythene to protect them from extreme frost, snow and damp.

Put straw round a few of the rhubarb crowns and then cover them with special rhubarb pots or a loose layer of black polythene, to bring on a few early sticks.

Dig the leeks, parsnips and turnips for kitchen use.

Protect sprouting broccoli and other greens from the ravages of the pigeons and pheasants, rabbits and hares. We all have our own method of doing this. We have found that 'Croptect' – lines of narrow strips of polythene tied tightly over the plants – is effective.

Peas and broad beans sown in November will now need looking at and protecting from frost and snow. We use either cloches or a covering of pruned twigs from deciduous shrubs such as *Artemisia abrotanum* and santolina, stuck in the ground along the rows. Mice will quickly wreak havoc, so it is wise to keep a strategically placed trap – but do cover it with an inverted seed tray or something similar to make sure you do not catch your favourite robin instead.

If you enjoy your own vegetables, polythene cloches put over part of a row of perpetual spinach or the last of the autumn lettuces will give you one extra meal.

Prune the autumn-fruiting raspberries, cutting the old canes to the ground.

January

The routine jobs will be the same as in December.

Spray the apple and pear trees with a winter wash as a precaution against aphids.

Remove all stumps and dig over the ground, but where you have cut the cabbages and left the stumps in the ground, small 'heads' will have sprouted, providing welcome greens when other fresh vegetables are scarce.

In an open, sunny position plant garlic cloves – you should already have planted a row in the autumn.

Put onion sets into seed trays by the end of the month so that in February or March (according to the weather), when the time comes to line them out in the garden, they will have enough root anchorage to prevent the birds from pecking them out.

Remember to attend to the indoor seed sowing of parsley, lettuce and cauliflower.

February

Two weeks before you intend to sow further rows of peas and beans, and the early carrots, warm up the soil by lining out simple polythene tent cloches or by putting clear perforated polythene flat on the ground.

By the end of the month the lettuces and cauliflowers sown in January may be put out under cloches after they have been hardened off in a cold frame.

Erecting supports for climbing beans always takes time, so if possible do this in February before spring work in the vegetable garden keeps you really busy. We use stout 2.5m (8ft) bamboos; if they are stored in a dry place in winter they will last for several years.

This is the moment to plant shallots and Jerusalem artichokes.

Prepare your seed bed, bringing the soil to a fine tilth.

You can, if necessary, cover the area where the first brassicas will be sown with clear polythene to help warm the soil. This will also encourage weeds to germinate, and they can then be removed before you sow the proper seeds. As soon as the annual weeds start to germinate, it is a sure sign for gardeners that their own chosen seeds can be sown outside and expected to germinate successfully. Sown too soon, they will lie dormant and cold in the soil, a prey to every passing predator.

By the end of February the vegetable garden, like the flower garden, is full of promise and anticipation. The soil is clean and manured, the flower buds on the fruit trees are swelling, and every day the soil is warming up, preparing for spring.

Above: One of the great joys of winter is watching the birds as they flock to the bird table. Many of the finches will come from the hedgerows to be fed once they have gained confidence. Sticks inserted horizontally at intervals into your feeder will give the birds something to perch on as they feed. Instead of nuts, you can fill the feeder with fat or bread for a change.

Right: The blue tit's plumage looks fresh and colourful against the snow-covered ground. The wonder of having a bird table is that it gives you the opportunity to observe the birds' behaviour and, after a while, some of them may become individual characters, particularly the robins and the tits.

Feeding the Birds

When I wake to a snow-covered world, my first job is to go out and clear the bird table, giving the birds some extra food and unfrozen water. Encouraging birds to come to your garden is a pleasure but nevertheless an important winter job. Getting to know them and their habits can become part of one's life.

In his Sussex garden, Christopher Lloyd enjoys the birds for their activity, character and song, while Allen Paterson, formerly at the Chelsea Physic Garden in London, takes pleasure in the colour they bring to his present garden in Hamilton, Ontario: 'Winter colour is brilliant and comes to the bird tray: cardinals made out of scarlet plush, evening grosbeaks like thrush-sized goldfinches, blue jays, red polls and purple finches.' Another gardener who keeps his bird table well stocked is Bob Dash. He sees the results with an artist's eye: 'Bright yellow corn meal in the bird feeders is a joy, as are the blue jays, chickadees, mourning doves. And then there comes a hawk and there is a kill and I see red.'

If you love to have birds about in your garden, especially in winter, there are many different ways to encourage them. The surest way is to have a feeding table, kept regularly supplied so that you never disappoint them. Make the food easy to get – the whole point is to help them to survive, not to test their ingenuity. Remember that water is as important to them as food: birds need to bathe to keep their insulating feathers in good condition. When it is freezing outside, a good way to ensure that they have a ready supply of water is to use an old frying pan or tin which you can bring indoors and thaw out on the stove. I do this every morning while the kettle is boiling for my breakfast coffee.

Bird feeding tables need thought. You want them where you will be able to see what is going on, partly for your own enjoyment and partly to make sure that it is the birds and not the neighbours' cats who are getting the benefit. You do not want the food to get sodden with rain or covered with snow, so some form of roofing is essential. If you have squirrels about, you will need to devise a means of thwarting them. A piece of wire netting or a tangle of thorns halfway up the post of the bird table is usually sufficient, and this will deter cats, too.

In a way birds are rather like children: you must build up their confidence. Start by giving them breadcrumbs and bird seed – foods which look natural. The blackbirds and tits will probably be the first to arrive as soon as natural food becomes

scarce. Blackbirds and thrushes seem to prefer feeding on the ground, so scatter some scraps, breadcrumbs and windfall apples, sliced in half, on the grass. The tits may be argumentative, or appear so – I often think they are chasing each other in sport rather than aggression.

Once you start on your routine of feeding, you will become aware of their likes and dislikes. I give them a varied diet, including baked potatoes cut in half, old pieces of cheese (grated if I have time), oatmeal, bacon rinds and the remains of cooked rice. If the birds like something different, they will get more. It may well be two or three days before they have the courage to try coconut if they have never had it before. Eventually they will, and then they will never have enough of it. Do not throw away the empty coconut shells – keep them,

and when you get extra fat from the butcher, run this down and tip it into the shells, hanging them up where you can see the tits at their antics.

The dunnocks – or hedge sparrows – come into the garden from the hedgerows in winter; in summer their diet consists of insects, and when these become scarce, they will search for seeds in the garden. It is worth saving and storing marrow and melon seeds in summer to give them a change of diet in winter. Wash the seeds to get rid of the pulp and dry them in a low oven before you store them away.

You can hang feeders filled with nuts and fat on wall brackets on nearby trees, on old stumps and on tree branches. Nuts, seeds and pieces of fat will wedge into the crevices of the bark. This is especially good for shy feeding birds like tree creepers, goldfinches and wrens. The gnarled and deeply furrowed trunk of our old acacia tree is marvellous for this; in summer the tree creepers spend a lot of feeding time there, presumably pulling out insects that are lodging in the cracks, so they already think of it as a larder.

A garden that is well supplied with berries, seed heads and fruit will attract lots of birds as long as you do not have an army of cats prowling around to scare them away. The fruits and berries which the birds enjoy most are those we have thought about in other chapters for their appeal to us in winter: cotoneasters, crab apples and sorbus, pyracanthus and hollies, not to mention the masses of rose hips which ripen and become palatable at different times.

The seed-eating birds will have a feast of delectable seeds from the herbaceous plants which we have left standing in the borders: phlomis, many of the grasses, monarda, rudbeckia and agapanthus. If you cut these down too soon you will be depriving the tits and finches of a major part of their diet and yourself of a lot of pleasure as you watch them poised, swinging on the last stems, having a final pick at the seeds.

The hips of the Rugosa roses will have ripened in the autumn, but the tougher hips on many of the ramblers become attractive to the birds only after Christmas, so you want to spread your offerings through the winter months.

Describing some of the evergreens at Winterthur, Delaware, the late Hal Bruce noticed what it meant for the birds when the seed heads open. 'In autumn a kind of premature snowstorm occurs when abundant cones of Carolina hemlocks (*Tsuga caroliniana*) discharge their pale winged seeds by the thousands. And through autumn, winter and early spring this place is a haven for rare grosbeaks, siskins and crossbills – birds that feed on the seeds of conifers and seldom venture far from the evergreen forests of the far north.'

Another gardener who enjoyed seeing birds descend on her garden was the late Elizabeth Lawrence, one of my favourite garden authors. She wrote mostly about her own garden in North Carolina, noticing that the birds there ate few of the berries they were supposed to like – maybe in her climate the ripening of the berries and the cold weather did not coincide.

In a Chicago garden I visited, a weeping cherry was the chosen place for special bird food, hung underneath the pendant branches, giving the birds a natural protection while they were feeding. The cardinals darting in and out were a colourful sight, and I could imagine how they would shine out on snow-covered days.

Birds need plenty of cover in winter; a well-planted garden is wonderful for them. I love to see the wrens flying out from the ivy on the walls and the tits and house sparrows perching on top of the clipped yews, flying inside them for protection on cold days. Before you trim the ivy too hard, spare a thought for the birds keeping warm and remember that it provides them with cover and shelter from the wind.

If you have a town garden, do not think you will get fewer, or less interesting, birds visiting you. You may well find quite the contrary – spraying in the country in summer keeps the number of insects down. By contrast, in towns there are likely to be more insects about in winter. Birds appreciate the warmth of towns, too: carrion crows perch on roof tops along with starlings and house sparrows.

One thing I am sure about: birds like humans to be regular in their habits. They are always waiting for their food to arrive. If you go away for several days, you really should ask a friend to come over and fill the bird table and containers for you, and see that they have water.

A Christmas Bunch of Flowers

'Anything that flowers from the second week in December till the end of January should find a place in every garden,' was the advice of that prolific writer of garden lore, the late E.A. Bowles. '*Iris unguicularis* is the best of all, and has often given me a large bunch of flowers for Christmas.' A bunch for Christmas Day itself is a challenge. Some years it is more of a contrivance, a conceit – as Tony Venison, gardening editor of *Country Life*, has rightly pointed out to me: 'It is the past year's weather that decides which flowers are open on Christmas Day'. He recounts: 'The dottiest tussie-mussie for a December 25th that has ever come within my experience was way back in 1977, when there had been no autumn frosts, only gales. Rain fell shortly before midnight on Christmas Eve, but Christmas Day itself was gloriously sunny.' His posy, of no fewer

As autumn yields to winter, berries appear, augmenting the scarcer flowers and foliage. This collage of winter colour, picked by Valerie Finnis, includes acer and eucalyptus leaves, viburnum, Iris foetidissima *seeds, hebe in leaf and flower, rue, mahonia,* Hedera helix *'Goldheart', variegated box, various rose hips,* Garrya elliptica, *coronilla, buddleja and violas.*

than fifty-nine blooms, is listed opposite on page 127. I wonder how any of us would score?

Are you as excited as I am on Christmas Day, taking a walk round the garden to discover what is in flower? The flowers I pick will go into a large, flat bowl of damp moss on the dining-room table. Inevitably, it is the Christmas rose, *Helleborus niger*, which excites me most. Look closely at the flowers and you will see how well nature has painted our pictures for us. Stiff sprigs of chimonanthus, the winter sweet, provide the essential ingredient of fragrance. Our bush is *Chimonanthus praecox (fragrans)*; each bloom on the bare branches has a purple centre. *C.p.* 'Luteus' has clear yellow flowers, opening a week or two later. Also for their scent, I will add a few sprigs of winter-flowering honeysuckle, *Lonicera fragrantissima*, and of witch hazel, hamamelis, which will be coming into bloom and producing a few precocious flowers.

Each year, of course, is different, but a shrub from which I have picked flowers on Christmas Day is, on rare occasions, *Chaenomeles japonica*. It grew in a sheltered corner in my garden some years ago, then had to be removed, but I feel inspired to try again – this time it will be *C. speciosa* 'Nivalis' for its pure white flowers.

The viburnums are proper standbys; they can be relied upon in almost every month of the year. There are four which are sure to give you flowers at Christmas. First, *Viburnum* x *bodnantense* 'Deben' and V. x b. 'Dawn'. I have the original *V.*

x *bodnantense*, planted years before the more exotic clones were available, and it has served me well – from October until March there is always a scented flower to pick for the house. Then there is *V. farreri* – once known as *V. fragrans* – given to my husband by a fine gardener nearby, to whom I am grateful not only in winter when it is in flower, but also in autumn when its leaves turn an alluring crimson-bronze before falling.

It was my son Christopher who pointed out to me the beauty of the evergreen *V. tinus*. I had given him and his wife some shrubs as a 'moving-in' present, among them roses with appropriate names for newly-marrieds: 'Wedding Day', 'Sweetheart' and 'Dearest'. They had spotted the viburnum in flower in December and asked for a shrub of this as well. Among the named varieties, *V.t.* 'Gwenllian' has rich pink buds, while *V.t.* 'Eve Price' makes a more compact bush and is better for the small garden. There is also a variety with a variegated leaf.

There are many other shrubs that will be opening in January and February, but I must keep to my Christmas bunch. The earliest of the prickly-leaved but sweet-smelling mahonias to bloom is *Mahonia lomariifolia*, whose lily-of-the-valley flowers start to open in November. The white flowers of *Prunus subhirtella* 'Autumnalis' will have appeared in time for Christmas, and *Choisya ternata* will often have a few white star-like scented blooms and the glossy, dark-green foliage is aromatic as well. Crush the leaves to release the pungent aroma.

Tony Venison's Christmas Bunch

Out-of-season and unexpected blooms

Cistus x *skanbergii* (an extraordinary beginning to any Christmas list)
Argyranthemum (*Chrysanthemum*) 'Mary Wootton'
Fremontodendron californicum
Alstroemeria psittacina
Gazania rigens 'Aureo-variegata' (with its petals open)
Gazania splendens 'Variegata'
Physostegia virginiana 'Vivid'
Linaria purpurea 'Canon Went'
Leucanthemum hosmariense
Olearia nummulariifolia
Bat's Double Red garden pink (raised before 1707)
Rosa chinensis 'Viridiflora'
Geum 'Lionel Cox'
Symphytum caucasicum
Cheiranthus 'Harpur Crewe'
Viola 'Nellie Britton'
Rosmarinus officinalis cultivars

Cuphea cyanea
Grindelia chiloensis
Lamium maculatum cultivars
Fuchsia microphylla
Primula vulgaris
Primula veris
Sundry primroses and polyanthus
Sundry auriculas
Various violets, pansies including 'Suttons Azure'
Ceanothus 'Delight'
Strawberry 'White Pine'
Bergenia 'Margery Fish'
Anemone 'De Caen'
Cyclamen coum

Characteristic winter-flowerers

Viburnum farreri
Viburnum x *bodnantense* 'Dawn'
Viburnum tinus
Viburnum tinus 'Eve Price'
Mahonia x *media* 'Winter Sun'
Mahonia x *media* 'Lionel Fortescue'

Erica x *darleyensis* 'Silberschmelze'
Jasminum nudiflorum
Daphne mezereum (by virtue of three individual florets)
Camellia oleifera
Arbutus unedo
Buddleja asiatica
Loropetalum chinense
Sarcococca humilis
Elaeagnus x *ebbingei*
Iris unguicularis
Iris unguicularis 'Mary Barnard'
Schizostylis coccinea 'Major'
Schizostylis coccinea 'Sunshine'
Schizostylis coccinea 'Mrs Hegarty'
Helleborus atrorubens
Helleborus olympicus
Helleborus lividus 'Boughton Beauty'
Crocus laevigatus 'Fontenayi'
Prunus subhirtella 'Autumnalis Rosea'
Prunus serrulata 'Fudan-zakura'

Far left: The delicate violet-blue flowers of Iris unguicularis *often appear before Christmas and flower right through to March. If you pick them when the flower stems have lengthened and the buds are still tightly-rolled umbrellas, you can watch them open indoors.*

Centre left: Helleborus niger, *the true Christmas rose, will usually provide a few early blooms for December 25th and will bloom on in profusion throughout January.*

Left: Noticeable for its elegant form, warmth of colouring and scent, Chimonanthus praecox, *the winter sweet, has upward arching stems carrying stiff pendulous golden blooms from December right through to February.*

In a sheltered corner, Galanthus x atkinsii *and* Lunaria annua *form an unusual combination. The outer cases of the seed heads of the honesty should be removed in autumn to show off the silvery 'pennies'. They will survive the ravages of autumn winds and rain if given a protected site, whereas the snowdrop thrives and flowers abundantly even when exposed to the severest of winter weather. Both have a fine delicacy of form that makes them complementary subjects for a Christmas bunch.*

Coronilla glauca flowers erratically, but I have had it blooming sparsely in December, when it was growing in a raised, south-facing bed. We have lost it there now and only have it in the cool greenhouse, where it flowers through the winter and looks much happier, so we leave it there. We do not have any camellias, but those who grow *Camellia sasanqua* and its cultivars should have a marvellously scented flower to add to their posy.

Some buds will be opening on *Daphne laureola*, but it seems a shame to pick them; I like to leave them for the berries, so that we will have seedlings appearing round about and young plants to distribute. But there are sure to be a few flowers to pick on some of the hebes, the most likely being *Hebe* 'Mrs Winder' and *H.* 'Autumn Glory'. We have found that the hebes with narrow leaves tend to be hardier than the broad-leaved forms, and 'Mrs Winder' has done well for us during recent hard winters. Her foliage is pretty, too, gradually turning from green to purple.

Regretfully we have no heathers in our garden, but the varieties of *Erica* x *darleyensis* which are lime-tolerant should be in flower, ranging from white through to lilac-pink.

There are always a few herbaceous plants flowering too, some by chance and some as regulars. The violet spikes of *Liriope muscari* may still be hanging on if you go searching for them among the stiff, grass-like leaves. I am surprised if there are no 'soldiers and sailors' – the common form of lungwort, *Pulmonaria officinalis* – opening their red and blue flowers. Winter pansies and violas are sure to be there, *Viola cornuta* still clambering through a nearby shrub and opening the occasional flower.

Christmas Day is early for snowdrops *en masse*, but *Galanthus plicatus byzantinus* should be showing some of its pure white flowers. E. A. Bowles, a great authority on snowdrops, wrote in 1915: 'G. byzantinus... were lovely through November and well into December and were especially pretty where they rose through a group of *Viola cornuta*.' A clump of the double *G. nivalis* 'Flore Pleno' is usually through on Christmas Day; nearby, *G. elwesii* is a well-suited companion to a low ivy. Some years flowers will have appeared; other years I have to wait another few days.

The most reliable early crocus is *Crocus laevigatus*, a lovely species with pale lavender feathered markings outside and a yellow throat and white anthers inside – a joy to study closely. *C. ancyrensis*, the attractive golden bunch crocus, may be showing but is more likely to wait until after the New Year.

The hardy *Cyclamen hederifolium* is sure to be over, but the plain green-leaved form, *C. coum*, should yield a few flowers if the weather has not been too unkind. I search among the vinca leaves and on the bed of polyanthus, too, to find a handful of blooms. We grow the lovely Barnhaven strain of polyanthus, which come in wonderful colours; if you get them into their winter homes early enough in the autumn they will have had time to resettle themselves and will carry a few flowers throughout the winter.

I do not have any roses flowering on Christmas Day, but remember envying Canon Ellacombe when he recorded in his classic work *In a Gloucestershire Garden* (1895), written from Bitton where he and his father were the incumbents for almost a century, 'I have picked a good bunch of fairy roses on Christmas Eve. I know no other flower that can come near the rose in this respect; the daisy is the nearest.'

By now I seem to have picked a large and varied bunch. They will not all be out every year, nor would I want them to be – half the excitement lies in discovering new and unexpected flowers each Christmas Day.

While I was assembling my own Christmas bunch, I thought how little I knew about what my transatlantic friends would be able to cull from their own, far more wintry, gardens, so I asked one plant lover – Ryan Gainey – to tell me about his choice, and one artist – Bob Dash – who I knew could be relied upon to produce some highly personal and exotic thoughts on a decoration for Christmas Day. Here are their 'bunches'.

Ryan Gainey writes: 'I took several walks and collected various plants from my own garden. From childhood I remembered decorating *Poncirus trifoliata*, with its great thorns, with gum drops, and I revived these memories by bringing in a good branch from a bush raised from seed several years ago. Hemlock, thuja and cedars provided the greenery for a wonderful wreath and the native American holly, *Ilex opaca*, a touch of red.

'I draped the chandeliers with great clusters of red nandina, looking like bunches of red winter grapes, and stuffed the inner boughs of the Christmas tree with dried honesty pods, to enjoy the glimmer when the tree was lit up. My mother brought her own offerings: great pine cones, their petals supporting a cluster of candies, and a wreath of grape vine studded with dried roses from her summer garden. The unrivalled sweetness of *Lonicera fragrantissima* drifted through the house, lodging forever a memory that that "sweet breath of spring" was always possible and always reliable.'

Bob Dash's Christmas offering from Madoo on Long Island is not so much a tussie-mussie, more a still-life – he never sets out to pick flowers, 'wishing neither to murder beauty, nor to bring death into the house'. Instead, with his highly personal approach to decoration, he chooses 'storm-wrack and prunings'. He makes 'Christmas massing with dried blooms, seed pods and the odd, graceful branch, erupting from, yet cushioned by, grasses', in much the same way as he 'salutes the other major occasions'.

As an artist, his preference is unsurprisingly for 'dried blooms that have slightly unfamiliar shapes, such as imploded cardoons, pronged teasels, spiky thistles and umbels of elephant garlic, for kitchen leeks gone to bloom in their second year and seedcases of *Magnolia grandiflora*, trumpet vine and wisteria'. His taste in grasses is correspondingly discriminating – *Miscanthus sinensis* 'Variegatus' and 'Zebrinus', marsh-culled phragmites, equisetum and *Cortaderia selloana*. His guiding factors are ever 'mass and shape', with 'benders and uprights for a mix of celebration'.

To mark Christmas in an especial way, he adds 'somewhere, pots of amaryllis (only the palest pink, please, definitely not those fire-engine reds, striped or picoteed) and forced narcissus, or a container of *Begonia rex* – if it is in full truss' – set in the midst of the still-life.

He has 'a fine standard indoor lime – *Sparmannia africana* – and if it is displaying, the whole arrangement will be on the floor, or in a corner, or circling the ivied post that supports the sleeping loft' in his winter house. For scent he has bowls of potpourri and says that 'there is nothing quite like lemons and limes rolling out from the abundance'. He is adamant that 'all of this is quite a lot better than holly, poinsettia, pine or spruce, all of which have gone tatty through over-use'.

I am more of a traditionalist – I cannot imagine Christmas without holly or hellebores – but I feel nothing but admiration for such an imaginative scheme of decoration.

PLANT PORTRAITS

I have chosen the plants in the following pages for the beauty of their winter bark, foliage and flowers and for the colour, scent, form and texture that they bring to the garden throughout the year. Plants have been zoned to guide American gardeners in their cultivation but, inevitably, in every zone there are differing local conditions which must be taken into account. Information on the system used is to be found on page 161.

Abeliophyllum distichum is a member of the olive family from Korea. It was exhibited at Vincent Square in the 1930s by John Coutts, then curator of Kew.

A small shrub, usually 90cm (36in) tall with the same spread, it has white scented star-shaped flowers, flushed with pink, which appear in February. It will do best against a warm wall, as it is growing at Wisley in Surrey, where it flowers freely, making a graceful, arching bush some 2.5m (8ft) tall.

I sometimes wonder why plants of merit do not become popular – in this case, could it be its awkward name? I believe it should be grown in more gardens, and it is definitely a plant which more nurseries should supply.
Deciduous shrub   Zones 4-9

Acacia dealbata, the common mimosa, is a native of Australia and Tasmania, and was introduced into Europe in 1820. On the French Riviera, where traditionally it flourishes, it is called the silver wattle.

The masses of fluffy fragrant yellow flowers emerge in winter and early spring, amidst finely divided, almost fernlike, silvery-green foliage.

There is a splendid specimen at Vauville near Cherbourg in France. It also grows successfully in the sheltered gardens of the south of England, reaching some 7.5m (25ft). At Dartington Hall in Devon it covers a whole wall. If your own climate is too cold and you decide to have it in your conservatory, remember that it will only take a modicum of lime and give it an acid soil.

In the USA mimosa is *Albizia julibrissin*, whose pink flowers do not open until July.
Evergreen shrub or small tree   Zone 9a

Viburnum tinus (see page 160)

Acer davidii is for me the best Japanese snake-bark maple, followed closely by *A. grosseri hersii. A. davidii* has three forms in cultivation. The first, collected by Charles Maries in 1877-8 for the great nursery firm of Veitch, is compact in habit; the second, a looser form with young stems of a striking purple, was sent to Britain from China later in the 19th century by George Forrest and bears his name; the third, which also commemorates its finder, is a more compact form discovered by E.H. Wilson in west Hubei and west Yunnan.

It is exciting to come upon one of these trees unexpectedly in woodland or as a feature planted at the end of a path. It is a good small tree – eventually reaching 9 to 12m (30 to 40ft) – with dark green summer leaves and strong autumn colour and fruit. The young bark is darkish green, turning later to purple-red and becoming finally striped with white, especially if it is planted in a shady position. Like all the Japanese maples, it thrives best on acid soil.
Deciduous tree   Zones 7a-9a

Acer griseum, the paperbark maple, was sent as seed to England in 1901 by E.H. Wilson on his first plant-hunting expedition to China. It grows up to 13.5m (45ft) tall. The largest and most striking specimen I can remember seeing is in the Arnold Arboretum in Boston – no coincidence since Wilson later joined the staff.

The grey underside of its leaflets gives this maple its specific name, but it is its brilliant red leaves and rich brown bark – which later peels off to reveal cinnamon colour underneath – that make it one of the best for October-November colour and winter interest. If you are planting this maple in your garden, choose one which branches from low down, to give you the full benefit of the paper-like patches of rich brown. This maple is all too scarce because its seed has a poor germination rate.
Deciduous tree   Zones 6-7b

Acer palmatum 'Sango-kaku' was first culti-vated in the West in the 1920s.

With its upright, generally narrow, habit of growth, it is an elegant tree at all times of the year, but in winter the sun shining through its

red branches is a truly wonderful sight. If you can spare a branch it looks spectacular in a vase indoors in winter. It is a tree I covet and, although difficult to find, I recommend you search for it. There is a huge specimen at Caerhays Castle in Cornwall.

It is hardy, even quite far north in America, and will eventually reach 9m (30ft), but it is slow-growing, even when given its preferred open site and rich, moist soil.
Deciduous tree   Zones (5b-9a)

Adiantum pedatum is more commonly known as maidenhair. Its name comes from the Greek word *adiantos*, meaning dry; and indeed the leaves of this family of ferns throw off any water that lands on them.

Truly one of the hardiest of ferns, its black, wiry stems and dying foliage are attractive among other low plants in the garden in winter. It will grow to about 45cm (18in) in good conditions – it needs shade and moist soil because its delicate leaves brown easily.
Deciduous fern   Zone 4-

Adonis amurensis was brought from Man-churia to the West in 1896. Named after one of Venus' lovers – Adonis – its specific name comes from the River Amur in Siberia, where it is also native.

It makes a spectacular carpet of colour in February and March, when the yellow, white, rose or red-striped flowers, 5cm (2in) across, stand just above the leaves, which are deeply divided into narrow, fern-like lobes.

Growing from a mass of cord-like roots, its preference is for light, loamy, acid soil. As its natural habitat is on north- and east-facing slopes with good drainage, it needs to be cool in summer and will flourish under deciduous trees.

Plant and divide in June just as the leaves are fading, and add fertilizer when the root system is growing, from autumn through until spring.
Herbaceous perennial   Zone 3-

Ajuga reptans 'Multicolor', in some cata-logues called *A.r.* 'Tortoiseshell', is native to temperate Europe and Asia and has a rich history in the literature of herbals as the bugle plant, a traditional herb for wounds.

The bronze leaves are attractively splashed with pink and cream, which set off the blue flowers later in the year. Not more than 12cm (5in) high in leaf, one plant will eventually spread to 25 or 30cm (10 or 12in).

All varieties of ajuga are indispensable as evergreen groundcover under trees, in woodland, in the mixed border and in the rock garden. I use *A.r.* 'Multicolor' in my double-tier planting schemes, with early bulbs – *Iris reticulata*, puschkinias, scillas and the lovely species crocus – pushing through it.
Evergreen groundcover   Zones 4-8

Alnus incana 'Aurea', a variety of the grey alder, is a member of a family honoured more in literature for its use once cut than for its qualities as a living tree. It is much slower-growing than its parent, *A. incana*, remaining a 6m (20ft) tree suitable for the average garden.

It is at its best in winter, when the previous summer's growth of wood becomes bright orange, and in spring, sometimes as early as February, when the red-tinted catkins elongate. It looks good against a background of dark leaves, such as yew or laurel. I have never forgotten seeing its bright winter wood in the Brooklyn Botanic Garden in New York.
Deciduous tree   Zone (3-)

Anemone blanda, the blue windflower, was introduced to the West from Asia Minor in the 1890s. It must be among the best-loved of the late winter flowers; every good gardening writer has sung its praises.

The brilliant blue of each flower is softened by an inner circle of white; equally prominent is each bright yellow eye. There are varieties – mauve, pink and white, and also a bright red with a white centre called 'Radar'. My favourite is *A.b.* 'White Splendour'.

I like the idea of a carpet of these anemones as groundcover, as bedding, or even in grass. They prefer half-shade, either in grass or in the crevices of rock, and appreciate a rich, well-drained soil.
Corm   Zones 7-9

Arbutus x andrachnoides is the hardiest, and in my opinion the most desirable of the strawberry trees – ornamental evergreens with glossy leaves. It is a hybrid of *A. andrachne*, which grows wild in Greece and flowers in March and April.

Its clusters of white flowers can start to open in December, sharing the branches with last season's strawberry-like fruits and rich, cinnamon-red peeling bark. Eventually but slowly it reaches 9m (30ft) and will tolerate lime.

To my regret, I have not succeeded in establishing *A. x andrachnoides* in my own garden. I have failed also with the variety *A. unedo*, which is more readily available in nurseries and sometimes, container-grown, in garden centres. In spite of these words of warning, I would recommend you to persist with the strawberry tree. You can see good specimens at Kew, Bodnant in Wales, Crathes Castle in Aberdeenshire and Highdown in West Sussex.
Evergreen tree   Zone 9a-

Arum italicum 'Pictum' has a family name given to it by Theophrastus – the Greek name *aron*, which records its European and Mediterranean origins. *A.i.* 'Pictum' is the exotic garden variety.

Its glossy green leaves appear in autumn – just as other plants are losing theirs – marbled or veined with ivory. As the leaves pierce through the soil, they unroll from their thick, indestructible tubers. They then disappear, to be replaced by stems of red berries, which flaunt themselves in late autumn and early winter. Attractive as they are, they are poisonous, so regretfully I remove them before they tempt the children with their seemingly harmless look.

They seed themselves in every odd corner of the garden, especially in shade. A good specimen can grow up to 40cm (16in) tall.
Tuberous rooted perennial   Zone (7b-)

Arundinaria viridi-striata – recently renamed *Pleioblastus viridi-striatus* – is an evergreen bamboo from Japan. It has been cultivated in Europe and Asia since 1870.

Its leaves, which are narrow and striped longitudinally with buttercup yellow, remain until late December; its canes are purplish-green. It grows to 90cm (36in) and spreads – remember that it is invasive, so either give it an area of its own or restrict its roots.

Adonis amurensis

Anemone blanda

Margery Fish, who knew it as *A. auricoma*, wrote: 'I grow it in the open and plant *Anaphalis triplinervis* in front of it.' Alternatively, as a foil for its yellow leaves, use deep red-leaved berberis or *Cotinus coggygria purpureus*. *A. murielae* (syn. *Sinundinaria murielae*) is a clump-forming species with arching stems. In autumn its bright green leaves take on a yellow tinge that lasts through the winter.
Evergreen bamboo   Zone 7-

Aucuba japonica 'Crotonifolia' is, I hope, coming back into fashion once again after its decline through association with dismal Victorian shrubberies. This species is the male form – if you want the advantage of berries, you should also have a female, *A.j.* 'Gold Dust'.

All the aucubas have handsome, glossy leaves and will flourish in almost any soil and in shady places. If you forget this and plant yours in a pleasant situation in the sun, it will become a

3m (10ft) specimen in its own right and you will be able to enjoy its golden-marked leaves outside or use branches for your winter vases. You can try planting *Taxus baccata* 'Repandens' round its feet.
Evergreen shrub   Zones (7b-9a)

Azara microphylla is a native of Chile and adjoining parts of Argentina.

It is at its best when grown against a warm wall, where it will reach a height of 3.6m (12ft). Each leaf is 2.5cm (1in) long and has a tiny stipule at its base, looking in fact like two leaves, one overlapping the other. Its small, mustard-yellow flowers are produced from February through until April.

I first saw this azara in a cottage garden, growing as a standard and looking beautiful with its spraying branches. I have planted it by the doorway leading into my garden and look for its flowers with anticipation towards the end

of each winter. Often, though, it is their vanilla-like scent which first makes me aware of them, especially on a warm day, when the flowers will waft their frangrance on the air.
Evergreen shrub   Zone 8a-

Berberis gagnepainii lanceifolia is a native of Sichuan, China, the seed collected by E.H. Wilson in 1904.

I like this shrub for its evergreen winter look, when the intensely prickly, narrow, and rather wavy leaves glisten in the sunlight. By December, when frost has done its work, some of these leaves turn red, and blue-black berries hang like jewels waiting to fall. Like others of the species, it has coloured shoots – in this case, they are a strong yellow. In summer its flowers are prodigious, appearing in June in bright yellow clusters – the bees become totally intoxicated!

I grow it as a free-standing shrub, 2.5m (8ft)

Arbutus x *andrachnoides*

Arundinaria murielae

tall. Alternatively, its very dense habit makes it an excellent choice for an impenetrable hedge. Cuttings are difficult to take and slow to root – seeds are the readiest means of propagation.
Evergreen shrub   Zones 6b-9a

Bergenia x schmidtii is the earliest of the bergenias to flower. Named after the 18th century botanist, Carl August von Bergen, the bergenias were regarded until quite recently as a sub-species of the saxifrage. Gertrude Jekyll, knowing them only as *Saxifraga megasea*, used them freely in her planting plans.

B. x *schmidtii* makes wonderful long-lived groundcover, with its broad, leathery, ever-green leaves, up to 23cm (9in) long and 15cm (6in) across, which turn an attractive purple-to-crimson in winter. Its bright rose-pink flowers, which open in January or early February, may be damaged by frost, but almost immediately new buds will come up and open.

Bergenias should be planted in bold groups, in the open, under trees or beside water. To make new plants you may pull off the fleshy stems at ground level in early autumn and bury them deeply, or propagate them by seed. *B. ciliata, B. cordifolia* and *B. crassifolia* are all hardy and widely available.
Evergreen groundcover

Betula albosinensis septentrionalis is a birch from west China, the seed collected by

E.H. Wilson in 1908. It grows up to 30m (100ft) and a large specimen tree, presumably grown from seeds brought back by Wilson, is still flourishing at the Arnold Arboretum in Boston; another may be seen at the Morris Arboretum in Pennsylvania. Others are to be found in Britain at Bodnant in Wales, Caerhays in Cornwall, Westonbirt in Gloucestershire and the Edinburgh Botanic Garden.

In Wilson's own words, 'the bark is orange-brown or orange to yellowish-orange or orange-grey... singularly beautiful and makes the tree conspicuous in the forest.' I think they look wonderful in a wild area of the garden where the trunk, lit by the winter sun, can be seen shining in the distance. On a visit to Comte Bernard de la Rochefoucauld's garden, Les Grandes Bruyères, near Orléans, I particularly remember the exciting trunks of the birches in the woodland among the heathers which give the garden its name.
Deciduous tree   Zone (5b-8a)

Betula utilis jacquemontii, discovered in the Himalayas early in the 19th century, was not cultivated in the West until the end of the century. It is the Western form of the Himalayan birch, *B. utilis*, whose bark can be light mahogany-brown instead of creamy-white.

The best of the white-barked birches, it can reach 9m (30ft) in height, and its stems are sometimes dazzling. Plant it where the sun will

catch it strongly in winter to lighten a shady corner, but remember to place it near the beginning of a glade, rather than as a focal point at the end where its brightness will attract the eye and make the glade seem shorter.

Give it other highly coloured stems as companions: red-stemmed cornus or willows will create a contrast you will enjoy all winter. Gentle annual scrubbing – enough to clean away dust and any green algae but not to peel off more bark than necessary – will bring out the colour. For the first few days afterwards, until it dries, the bark will have a pink tinge.
Deciduous tree   Zone 6

Buddleja auriculata is the South African member of that diverse race, the buddlejas, named after the 17th-century botanist, the Rev. Adam Buddle, vicar of Fambridge in Essex.

Its highly scented, creamy-white flowers, each with a yellow eye, bloom bravely from September through to January. It grows from 1.8 to 2.75m (6 to 9ft) high, and looks best against a wall. It is slightly tender, but has survived for many years against a sunny, sheltered wall at Kew and has done well as far north as Northumberland. There is a magnificent specimen on the wall of the Long Garden at Cliveden, Buckinghamshire. When I have a cool conservatory, it will certainly be one of my first choices.

Like other buddlejas, cuttings root easily, so

Betula utilis jacquemontii

Callicarpa bodinieri giraldii

Bergenia x schmidtii

keep your prunings for plants to give to friends. Other species to flower in winter are *B. madagascariensis, B.* x *lewisiana* and *B. asiatica,* which rivals *B. auriculata* for scent and which, despite its name, is very much a greenhouse plant.
Evergreen shrub   Zone 9

Buxus sempervirens 'Handsworthensis' is a 19th-century version of an ancient tree, the common box.

Its thick and leathery dark green leaves are large and broad, and it grows vigorously and upright. Like all box, this variety clips well, but will need to be kept in shape by regular attention. It is one of the best types of box for a tallish hedge, reaching a height of 1.8 to 2.5m (6 to 8ft). Box is also much used for topiary, though it does not lend itself to complicated shapes – stick to buns, squares or pyramids.
Evergreen shrub   Zones (6a-8b)

Callicarpa bodinieri giraldii, as its name suggests – *kallos* is Greek for beauty and *carpos* for fruit – is outstanding for its berries. Originally from China, it came into cultivation in the West at the turn of the century.

Its fruits are wonderfully colourful, ranging from a pale bluish-lilac to a rich purple; each berry is small but in bunches they make a dramatic display.

It will probably take two seasons to establish, making a mass of roots before it puts on much top growth and rewards its plants with fruit. Give it a good rich soil and a sunny position. For cross-fertilization and better berrying, plant several seedlings together
Deciduous shrub   Zones 6-8

Calluna vulgaris, commonly known as Scottish heather, used to be tied together in bundles at the end of a stick to make brooms. Hence the plant's name – *kalluno* means to clean in Greek.

Although heathers do not flower in winter, their foliage is so good that they are almost worth planting for that alone. For golden winter foliage, choose 'Beoley Gold' or 'Gold Haze'. Both have white flowers in September and grow around 60cm (24in) tall. 'Golden Carpet' is prostrate in habit and its foliage turns orange and red in winter, while 'Wickwar Flame', raised by George Osmond in Gloucestershire, turns bright flame-red. Two excellent greys are 'Silver Knight' and 'Silver Queen'.

They all need poor acid soil – rich soil will make them grow too well.
Deciduous shrub   Zone 5-

Camellia reticulata 'Captain Rawes'

Camellia reticulata 'Captain Rawes', an import from China, was the first form of *C. reticulata* to reach the West.

Its flowers are immense, often 15cm (6in) across, with wavy, rich crimson petals. Its tough, veined, toothed, shining green leaves are lovely at all times of the year.

Tolerably hardy, it is best planted against a west- or north-facing wall so that the buds are not induced to open too early in the year or to thaw too quickly after a frost. You can see a magnificent example of this winter-flowering camellia, 6 m (20 ft) high, in the Savill Gardens at Windsor Great Park in Berkshire. Like other camellias, it prefers an acid soil.
Evergreen shrub   Zone (9a)

Camellia sasanqua 'Narumi-gata' is one of the numerous dual-nationality camellias – native to both China and Japan. It was first introduced into Europe in 1896.

Its evergreen leaves are small and shiny; its flowers creamy-white, quite large and flushed with pink towards the margins. The first blooms open well before Christmas and continue through to February. Unlike most of the camellias, the flowers are scented.

A tall plant, it can reach 6m (20ft) and spread to 3m (10ft), but it is not reliably hardy in Britain. In all but the most sheltered gardens, it is best grown in a sunny position against a sheltered wall.
Evergreen shrub   Zones (8b-9a)

Chaenomeles speciosa 'Simonii'

Chaenomeles x superba 'Rowallane' has parents which come from China and Japan, although its common name, japonica, claims only one nationality for it.

The large, blood-red flowers open from late autumn onwards through to February, depending on the weather. It is one of the winter standbys for blossom, which appears on the old wood, so the young shoots with their newly grown leaves should be cut back firmly.

Very hardy and suited to most soils, the plant is a hybrid of the garden quince – it produces typical 'quinces' which can be made into jelly. Reaching 1.8m (6ft) in height and spread, a well-trained specimen on a cottage wall can be an eye-catcher, but will require careful pruning to achieve such an effect.

If I had space I would choose *C. speciosa* 'Moerloosii', with its delicate mixture of pink and white flowers, and *C.s.* 'Nivalis', with plain white flowers. *C.s.*'Simonii', with its bright red flowers, has a spreading habit.
Deciduous shrub   Zones 5-9a

Chamaecyparis lawsoniana 'Lanei' is one of those rare plant names that is absolutely explicit. The Greek word *chamai*, meaning on the ground, tells us how it grows; *kuparisson*, that it is a cypress; Lawson that it was introduced by the Edinburgh nursery in 1854, from Oregon; and Lane, that it is a strain raised by the Berkhamsted nursery in 1938. The only information missing is acknowledgment that it is golden, although some nurseries add 'Aurea'.

Its golden-yellow foliage stands out brightly in winter against darker backgrounds – the leaf sprays having greener undersides.

Inspired by Gertrude Jekyll's writings, we made a border of predominantly yellow flowers and leaves, in which the columnar chamaecyparis, now 4.5m (15ft) tall, provides a solid background, itself framed by a dark yew hedge.
Evergreen tree   Zones (6-8a)

Chamaecyparis pisifera 'Filifera Aurea' is a form of the Sawara cypress from Japan.

A small- to medium-sized conical tree or a large-domed shrub, it has spreading branches with leading shoots that droop. Rounded in shape and with dense foliage, it can reach a height of 4.5m (15ft) or more, while it grows. Therefore, allow it enough room, using short-lived plants to cover the ground around it while it expands.

Try using it in front of or beside a dark-leaved holly, or as a foil for a golden holly such as *Ilex* x *altaclarensis* 'Golden King'. A pair of these golden cypress look good flanking a gateway; a single specimen used in a shady corner will also work well.
Evergreen tree or shrub   Zones (5a-8a)

Chimonanthus praecox has the most expressive name of all for a book on winter gardening – in Greek, *cheimon* means winter, *anthos* a flower. The Chinese counterpart of the North American allspice, *Calycanthus floridus*, its common name in Britain is winter sweet. It was introduced from China as early as 1766.

It is outstanding as a winter-flowering shrub with round buds which open to reveal starry, yellow flowers with central rings of shorter purple petals. It should be grown against a wall near the house, so that its fragrance of violets and jonquils can filter indoors. In summer its leaves unfortunately are undistinguished, so give it a climber to enliven it – clematis, sweet peas or morning glory.

Remember Vita Sackville-West's verdict: 'It will not begin to flower until it is five or six years old. But it is worth waiting for.'
Deciduous shrub   Zones 7-9

Chionodoxa luciliae combines two Greek words – *chion* meaning snow and *doxa* glory of – and glory of the snow is what it is commonly called. George Maw, a famous gardener of the mid-19th century, gathered up a collection of chionodoxa bulbs on the summit of Mount Nif on a visit to Turkey in May 1877. He grew them successfully in his garden in Shropshire and distributed them among his friends.

Its large flowers, some 3cm (1 1/4 in) across, are of an exquisite pale sky-blue, shading imperceptibly to white in the centre. The most common and easily grown of this small genus of very early spring bulbs, they ought to be planted with scillas – close relations – and grape hyacinths in a mass. In the wild they flower at the melting snow line, so if you wish to see them in Turkey on Nif or Boz Dagh, time your visit well.
Bulb   Zones 4-8

Choisya ternata, the Mexican orange blossom, takes its name from the 19th-century Swiss botanist, Jacques Denis Choisy.

Reaching 1.8m (6ft) in height, it is a splendid shrub to plant near the house, with its round, shapely form and glossy, dark green leaves, which release a pungent aroma when crushed. Its scented flowers appear sometimes in midsummer, sometimes in winter.

For many years choisya was thought of solely as a glasshouse shrub, yet William Robinson records seeing it growing outside. Earlier this century, a Fulham nursery offered it as a mushroom-shaped standard. A line of them alternating with hibiscus could cause quite a sensation if they flowered simultaneously. For the golden border there is a new yellow-leaved form, *C.t.* 'Sundance'.
Evergreen shrub   Zones 8b-9a

Chrysanthemum parthenium aureum, the golden feverfew – now known as *Tanacetum parthenium aureum*, is the golden variety of the old physic herb grown widely as an antidote to fever and headaches. It is native to Europe and reliably hardy.

I use this perennial herb, 30 to 38cm (12 to 15in) high and with delicate golden-green foliage, as an edging in full sun or as a bright element under dark yew. I like to grow it in odd corners, on walls and under hedges. It will seed itself freely if allowed to flower, but the foliage remains in better condition if you remove the flower buds before they develop.
Herbaceous perennial   Zone 5-

Clematis armandii was sent to Britain as seed from China in 1901 by E.H. Wilson. A later batch was sent to the Arnold Arboretum in Boston.

It is one of the evergreen climbers I rate very highly for its winter leaf and early spring flowers. The variety 'Apple Blossom' has evergreen, glossy leaves and bronze tendrils studded with pink flowers.

It will reach at least 3.6m (12ft) in height – even more when it is well-situated. Every reference I have read recommends that it should have a place on a warm house wall, facing south. At La Mortola on the Italian Riviera it almost covered the huge roof of the summer house at one time. Further north, it is not reliably hardy, being killed to the ground during very hard winters, but it usually resurrects itself. Its deadheads are unattractive, so plant it where you can snip these off without too much ladder work.
Evergreen climber   Zones 7b-9a

Clematis cirrhosa balearica, as the reference to the Balearic Islands suggests, comes from southern Europe and Asia Minor, and has been grown in England for 400 years.

The bell-shaped flowers of this evergreen clematis are scented and creamy-white; its delicate fern-like leaves are doubly attractive in winter when they become tinged with bronze. Reaching 3m (10ft) in height, it should be planted in a good position where its flowers, which come in January to March, can be seen easily. It requires a south or west wall.
Evergreen climber   Zone 8

***Cornus alba* 'Sibirica'** is commonly called the Westonbirt dogwood, having been planted by George Holford at his Gloucestershire arboretum from a strain growing wild in Siberia.

It is the brilliant, sealing-wax red haze created by the stems of this cornus that are so wonderful in winter, even though its name suggests that the white berries are its chief glory. Plant them in thickets, not as single specimens, and prune them hard, either every spring or on alternate years, to achieve the young growth which provides the most vivid red stems we enjoy in winter.

I have made a compromise and planted *C.a.* 'Spaethii' for its golden-margined leaves in summer and its not-quite-so-bright red stems in winter. Early flowering bulbs – crocus and puschkinia – planted at its base add interest.

The flowering dogwood of North America, *C. florida*, has the most spectacular fall colours.
Deciduous shrub   Zones 3-8

Cornus mas, the Cornelian cherry, retains its original name – cornus. Linnaeus used this name to embrace all the dogwoods or cornels.

In February its tight, round buds, the size of large pinheads, expand and open into acid yellow flowers with yellow-green bracts, lasting in their glory for weeks – those weeks when I am anticipating spring yet enjoying winter. Its fruits are red and cherry-like. Its summer foliage, like that of many winter flowerers, is unimpressive.

It needs space as it will make a handsome spread of at least 5.5m (18ft). It does particularly well on limestone soil but, in common with many of the dogwoods, it may need a south wall if it is to flourish.
Deciduous shrub or small tree   Zones 5a-8a

***Cornus stolonifera* 'Flaviramea'** is a variety introduced into England by Spaeth in 1899. The American Indians used the bark of *C. stolonifera* – known in the USA as *C. sericea* – to make kinnikinic, a subsitute for tobacco.

Clematis cirrhosa balearica

The young shoots range from yellow to olive green. It is perhaps not as exciting as its red-stemmed cousins, but planted in large groups beside water, its stems lit by the sun, it can look lovely. It makes a good contrast when planted with *C. alba* 'Sibirica'.

It needs little encouragement to sucker – the red-stemmed varieties tend to keep to their own patch – but does best in moist conditions. Once established, it should be cut, annually in spring or on alternate years, right to the base to encourage new young stems to shoot in time for the winter colour they bring.
Deciduous shrub   Zones 3-8

Cortaderia selloana, better known as pampas grass, came to Europe from South America in 1848.

The upright, silvery plumes of flowers – 3m (10ft) high – persist well into winter and have a gentle and elegant habit – the female spikes being more beautiful than those of the male. The narrow leaves are very tough and sharp and can inflict unpleasant cuts.

To increase pampas, divide the existing clump in April just before the new growth starts. The most effective way to control and tidy old clumps is to set a match to them every 2 or 3 years in early spring to burn off the dry leaves.
Perennial evergreen grass   Zone 8-

Corylopsis pauciflora was introduced into England from Japan in 1874 by the London nursery, James Veitch.

One of the most beautiful of early-flowering shrubs, its delicate racemes of pale primrose-yellow flowers have a cowslip scent. In a mild winter, the first of the flowers, borne before the leaves, come into bloom by the end of February.

Reaching 1.8m (6ft) in height and spread, it is happiest on acid soil, cannot abide chalk and is not keen on lime. It looks wonderful carpeted underneath with chionodoxas or *Crocus tommasinianus*, especially in dappled woodland sunlight.
Deciduous shrub   *Zone 6a-9a*

***Corylus avellana* 'Contorta'** bears the Greek name for a hazel nut – *korylos*. Its curiously twisted habit of growth is suggested by the Latin word *contorta*; its common name is the corkscrew hazel.

Native to the British Isles, it is slower growing than the other hazels but will eventually reach 2.5m (8ft). Its long, lambs' tail, male catkins appear in February.

All dedicated flower arrangers should have this hazel in their garden for its amazing shape. Cut a bare branch in winter and it will stay in good condition for weeks – in fact if you stow it away during the summer, you can bring it out again next winter.
Deciduous shrub or tree   *Zones 3-8*

Corylus avellana 'Contorta'

Cotoneaster distichus tongolensis is one of the vast clan of cotoneasters and one of the best of the dwarf varieties. In Latin, cotoneaster literally means inferior quince, but you cannot always trust plant names.

For a slow-growing, medium-sized shrub for the border, this is definitely my choice from among the innumerable berrying cotoneasters, especially if it can be seen from the windows. Its bright scarlet berries are outstanding for their colour and size, and have the great virtue of being unattractive to birds. The leaves, which are slightly woolly, hang on until late winter.

There is a cotoneaster of every shape and size for every corner of the garden – deciduous or evergreen, upright or spreading. They are all tolerant of almost any soil, hardy in almost any climate and berry well, so you should choose yours for its shape and size. If you want a yellow-berried form, *C.* 'Rothschildianus' is the best.
Deciduous shrub   *Zone 5*

***Crataegus monogyna* 'Biflora'**, a form of the common hawthorn, is that legendary tree, the Glastonbury thorn. Its name comes from the Greek word *kratos*, meaning strength. The Glastonbury legend has it that Joseph of Arimathea thrust his hawthorn staff into the earth, whereupon it immediately 'grew, and constantly budded and blowed upon Christmas Day'.

The white flowers, which smell delicious, are the chief attraction in winter of this lean, spare and prickly tree.
Deciduous tree   *Zones (5a-7b)*

Crocus chrysanthus grows wild, especially on stony hillsides, in Greece, Turkey, Bulgaria and Yugoslavia. These wild forms are usually bright yellow and sweetly scented, the flowers and leaves developing simultaneously. The many cultivars now available to us come in a wide range of colours – yellows, blues and white – and some are striped or feathered outside.

They make a wonderful show on raised beds in full sun, or you can plant them in pots, leaving them outside until they begin to show colour and then bringing them indoors to enjoy. Unfortunately the mice, voles and squirrels enjoy them too; unless you keep mousetraps nearby you can lose a whole colony of them in a few nights.

They should be planted in the autumn about 5cm (2in) deep. If they would prefer to be further below ground, they are able to pull themselves downwards with their retractile roots.
Corm   *Zone 4*

Crocus laevigatus fontenayi, when planted in a mass to flower from November through to January, gives an overall effect of lilac, with the outside of the petals feathered a rich purple. If you peer inside the flowers as they open, you will find a vivid yellow tinged with orange; bend down and their scent will delight you.
Corm   *Zones 6-9*

Crocus tommasinianus, which flowers in January and February, comes from western Yugoslavia, Turkey and Greece. It is hardy throughout most of the United States where it is slowly becoming more popular. There are wonderful drifts at Winterthur, Delaware.

This is the 'silver-lilac' crocus, looking sometimes almost grey. Variations of colour do occur when they are naturalized and allowed to grow freely in beds and borders and in the rock garden. They spread well when their seeds are able to drop straight into the soil unhampered by grass.

Plant some of the corms among the pretty, marbled leaves of the autumn-flowering *Cyclamen hederifolium*. The low *Narcissus* 'Peeping Tom', *N. cyclamineus* and *Fritillaria verticillata* also make good companions.
Corm   *Zones 4-8*

***Cryptomeria japonica* 'Elegans'** came from Japan in 1861.

It is a beautiful evergreen conifer, whose soft foliage turns from green to bronze and rich purple in winter. The trunk is reddish-brown and can bend over, especially when weighed down with snow; an unobtrusive stake will help. It does not grow nearly as tall as its 15m (50ft) parent, reaching only 4.5m (15ft), and so is suitable for the smaller garden.

I have planted it in my garden and enjoyed it, although on our alkaline soil it does not stay as elegant as on acid soil, where it will remain billowing and symmetrical for years. The handbook *Trees for the Landscape*, produced by the University of Georgia, recommends cryptomeria, adding that it prefers a moist soil, being only moderately adaptable to a dry site; it must then tolerate the hot Georgian summers.
Evergreen tree   *Zones 7-9a*

***Cupressus glabra* 'Pyramidalis'** grows quickly into a tall pyramid. It has small, blue-grey leaves of a wonderful texture and smooth, purple bark which peels away to reveal yellow underneath.

Reaching 10.5m (35ft) in height, its colour contrasts well with dark green conifers and makes a good screen where Lawson cypresses

might seem dingy. It also looks effective clipped as a hedge along a drive.
Evergreen tree   Zone 6-

Cyclamen coum is one of the most free-flowering and easily established of the hardy cyclamen. Growing wild in Greece and other countries of the Mediterranean, the cyclamen derives its family name from the Greek *kyklos*, meaning circular – a reference to the way the flower stems twist into spirals as the seed cases develop.

With thought, you can have cyclamen flowering almost all through the year, but for a grand display I would choose *C. hederifolium* to flower in August and September, followed by *C. coum* to flower through the winter. Its 10cm (4in) carmine-to-pink-to-white flowers often have a sweet scent and look beautiful emerging through a sea of mottled grey and green leaves, planted as sweeps or surrounding the base of trees. What better companion for early snowdrops?

Cyclamen are self-propagating but, if you want to sow seeds in another part of the garden, remember that they will not germinate until just before the flowers of the next season appear. Their corms will grow in girth annually, but only if they are established where they can remain undisturbed for years. If you buy them with leaves already growing you will have no problem planting them the right way up, but if they are dormant you will have to take more care.
Corm   Zones 6-9

Daphne blagayana was discovered by Count Blagay on his estate in Slovenia in 1857.

By the end of February, each stem of this prostrate evergreen bears a terminal cluster of sweetly scented, ivory-coloured flowers. It is one of the most desirable of all the daphnes, and I envy anyone who has a flourishing patch in their garden.

Preferring limy soil and partial shade, it needs an area where it can colonize at its own pace. The stems easily become bare and must be mulched with soil or layered with stones, under which it will burrow, rooting itself as it spreads. Some gardeners claim to throw a stone at it whenever they pass the plant, to encourage its growth, but it is not a hit-or-miss plant. It deserves the best care and attention and plenty of leaf mould.
Evergeen shrub   Zone 6-

Daphne laureola has a particular charm belying its common name, the spurge laurel.

Its yellow-green tubular blooms, which appear in February and March, are evening-scented, and its leaves are green and shiny. *D.l. philippi* is a recommended form.

Daphnes do not like being disturbed, so take care when you fork or dig round their roots. They need moisture and coolness for their root run and so prefer to be planted in shade where they will reach a height of 90cm (36in) and a spread of about 1.2m (4ft).
Evergreen shrub   Zone 7

Daphne mezereum, the paradise plant, has been cultivated in Britain for centuries and is recorded growing wild in Hampshire woodland in 1752.

Its fragrant flowers, ranging from white to purple, appear in dense clusters from February through to April and sometimes as early as January. The berries on the purple-flowered form are brilliant red; on the more unusual white form they are amber-coloured. Both are highly poisonous.

It is thought of as a cottage garden plant, but is worthy of a place in even the grandest garden, making a 60cm (24in) bush with very upright stems.

Not a long-lived plant, you will fortunately often find seedlings around it which you should nurture and keep as replacements.
Deciduous shrub   Zones 3b-7b

Dianthus caryophyllus is a plant of the gods – in Greek *di* is the word for Zeus and *anthos* for a flower. We know it, more prosaically, as a pink, after the blush colour of its summer flowers; in winter we enjoy its bright grey-green foliage.

The pinks are a vast tribe and few plants are more traditional to the English garden. They are easy to grow and nearly all are scented. Sun-loving, they need a well-drained soil, preferring it to be alkaline, but they are too good-natured to object to almost any save water-logged soil. If they have a fault, it is a tendency to flower themselves to exhaustion. Watch out for this, and as Vita Sackville-West advises, 'cut the wealth of flowers hard back to the grey-green clumps, to protect and save them from their own extravagant generosity'. If you take cuttings each August, this will ensure a good supply.
Evergreen herbaceous perennial   Zone 8

Elaeagnus pungens 'Maculata' was raised by Veitch's nursery and gained a first-class certificate in 1891. Its parent was brought from Japan in 1830.

Its leathery green leaves have a lustrous upper surface; each one is dashed along the centre with shining yellow, making it a spectacular plant, essential for a golden border. Its only defect is its slightly awkward and angular way of branching.

Allow it plenty of space to make a height and spread of 4.5m (15ft) . It has a tendency to revert – a shoot will suddenly produce pure green leaves, so cut them out at once.

Crocus chrysanthus

Elaeagnus pungens 'Maculata'

Erica carnea 'Springwood White'

My preference is for this form, but others think *E.p.* 'Dicksonii' and *E.* x *ebbingei* 'Gilt Edge', both with yellow variegation on the leaf edge, superior in colour.
Evergreen shrub   Zones (7a-9a)

Epimedium perralderianum, the most handsome of the barrenworts, was introduced into England from Algeria in 1867.

It has bright yellow flowers and distinctively toothed, trifoliate leaves. It is the only true evergreen epimedium; the foliage of the others should be cut back in February to reveal the flowers and delicate young leaves which push through at this time.

All the epimediums are clump-forming perennials, spreading slowly but surely by means of intertwined rhizomes to make excellent groundcover. Their foliage is beautiful all through the year, tinted in autumn and winter with shades of brown. They will tolerate dry shade and full sun but are happiest in deep rich soil and light shade. At Winterthur, Delaware, they are valued for their hardiness, their grace and delicacy, and their gifts of foliage and flowers.
Evergreen perennial groundcover   Zones 5-9

Eranthis hyemalis is known to us as the winter aconite, although the Greek name is the spring flower (*er* meaning spring, *anthos* a flower). Native to Europe, it is hardy everywhere in Britain and in the USA, except in the far north.

Once established, this tuberous-rooted winter beauty never fails. It flowers in profusion, forming a bright yellow carpet at the beginning of February – sometimes even earlier. The flowers open in sunshine, so do not hide them in shade. Though only 15cm (6in) tall in flower,

the leaves grow taller than that and form a dense mass in spring.

It likes a well-drained and reasonably fertile soil, preferably under deciduous trees and shrubs where it may remain undisturbed. It is best increased by seed, which ripens 3 months after flowering and is easy to gather and scatter. The following year typical rosettes of leaves will appear and from then on you can expect an increasing display of flowers.
Tuberous-rooted perennial   Zones 4-7

Erica carnea, the mountain heath, is also known as *E. herbacea*.

The plant is quite sturdy and bushy, often 23cm (9in) high. In mild winters it flowers in December or January, and a drift of its flowers will gradually take on a rosy-pink glow. It is one of the few lime-tolerant heathers. Ideally it should be planted in drifts, or used as groundcover for other, taller shrubs.

E. x *darleyensis* is a popular hybrid of *E. carnea* and *E. erigena* (syn. *E. mediterranea*). It has pale pinkish-purple spikes of flowers that bloom continuously from November through to March.
Evergreen shrub   Zones 6b-7b

Eucalyptus niphophila is an evergreen tree commonly known as the snow gum for the unusually white stems and trunk. Its home is the mountains of Victoria and New South Wales, but it has been grown successfully in Britain and is hardy in southern England.

The bark is especially striking as it exfoliates to reveal a snake-bark skin patterned with cream, russet grey and green. The leaves too are beautiful, catching the sun as they hang from their silvery stems – the colour is created by the bloom on the branches.

They should be planted as seedlings, ideally no more than 15cm (6in) high. Most are quick growers and will reach 9m (30ft) eventually. Do not stake this eucalyptus early in life, or it will rely on support for ever.

If you are intending to plant eucalyptus for its juvenile foliage, you should choose *E. gunnii*. Pollard the stems each year just before the sap starts to rise, and you will have good picking material all through the winter.
Evergreen tree   Zone 8-

Euonymus fortunei '**Variegatus**' is a sport of the green-leaved *E.f. radicans*, which was introduced from Japan in 1860.

The leaves of this shrub are slightly greyish and margined with white and yellow. When the temperature drops they become tinged with

pink, almost crimson at the edges. The colour is most vivid in spring when new leaves grow, pushing off some of the previous year's foliage. The fruits are typical spindle berries, round and cream-coloured, opening eventually to reveal lovely, tiny, orange seeds.

It is a hardy evergreen which will climb if given a wall or tree for encouragement; alternatively it will remain in shrub form, about 60cm (24in) tall if grown free-standing. The shrubs in my garden all came from one original specimen and I have used them in both ways, but it is the climbing shrub, growing up an old pear tree that I value most for winter effect, glowing all day in the slanting rays of the sun.

All forms of *E. fortunei* will make useful dwarf hedges and may be kept neatly clipped. In the eastern states of the USA where ivy is not hardy, *E.f. radicans* is often used to cover walls.
Evergreen shrub   Zones 5-9a

Euphorbia characias wulfenii is one species in a genus named after Euphorbus, physician to King Juba of Mauritania. It is a near-hardy perennial, native to the Mediterranean.

Blue-grey, somewhat hairy leaves are carried in spirals up the stems, which grow one year and flower the next. I enjoy watching the flower trusses develop through the late winter; the flowering stems nod their heads after the first frost, then raise them and open in late spring. Each flower is chartreuse-coloured; the species *E. characias* is similar but has a dark eye to its flower.

A shrub-like plant, it makes a weighty 1.2m (4ft) clump, most useful in a corner where the soil is well drained and an imposing plant is needed. Like other euphorbias, it will sow itself. Look out for seedlings, but do not transplant them unless you have to – they have awkward, fleshy tap roots and the stems, when cut, have a thick milky sap which can irritate the skin.
Herbaceous perennial   Zone 8

Euphorbia myrsinites has been known in Britain since 1570 and is a native of southern Europe. In their book on perennials Pamela Harper and Frederick McGourty recommend several varieties available to gardeners in the USA.

Its fleshy, blue-grey leaves grow in spirals along its prostrate, snake-like stems, which are never higher than 15cm (6in). The yellow flowers are unimportant – it is the foliage which is attractive throughout the year.

Though completely hardy in England, it is a short-lived plant, but characteristically generous

Eucalyptus niphophila

Festuca glauca

in the seeds it distributes around itself. Grown on a raised bed or with sharp drainage, it will self-sow abundantly.
Herbaceous perennial   Zone 5-

Fatsia japonica is a striking shrub, which should be used judiciously, perhaps as a focal point or in association with architectural features.

Hardy through most of the British Isles, it is quite a fast grower, eventually reaching 4.5m (15ft). It has large, glossy, deeply divided leaves – 38cm (15 in) across by 25cm (10 in) long – of a rich green with boldly marked pale veins. In October it becomes heavily laden with panicles of creamy flowers at the end of every stem. These develop later into black berries that linger through the winter. In common with many other evergreens, the leaves become limp and appear black after a heavy frost, but they soon recover as the temperature rises.
Evergreen shrub   Zones 8a-9a

Festuca glauca (syn. *F. ovina glauca*) is sometimes called the 'blue' grass.

It is a low, densely tufted perennial grass growing to about 15cm (6in). It is evergreen, trouble-free and improves in drought conditions, when the colour will intensify. It is

valuable both as a group in front of the border and as an edging, and will do best when divided frequently, probably once in spring and again in autumn. If you use it as an edging, you should cut away the flower spikes to help maintain a supply of new leaves. There are many kinds of festuca under the names *F. caesia, F. cinerea, F. glauca* and *F. ovina*.
Perennial grass   Zones 4-9

Fragaria x ananassa 'Variegata' is a form of the wild strawberry – the Latin name for strawberry is *fraga*. Although it has fragrant, edible fruit, these are seldom produced, and while it flowers in moderation, it is chiefly attractive for its variegated leaves of green-splashed cream.

It lasts the whole winter and I use it in the vegetable garden around the base of the trained apple trees to provide unusual groundcover. It can spread almost too fast with its many runners, but it is good at suppressing weeds and grows equally well in sun or shade.

Other uncommon strawberries are *F. vesca monophylla,* first noticed by John Tradescant in the 17th century, which has one leaf instead of three; and the Plymouth strawberry whose fruits are studded with tiny leaves instead of pips.
Evergreen perennial

Galanthus nivalis 'Flore Pleno' takes its generic name from the Greek*, gala* meaning milk and *anthos* a flower.

I have chosen this double form of the common snowdrop as it is the variety which has become predominant in my garden. It is sweetly scented, lasts especially well in water, and is excellent for naturalizing.

The three outer segments of the flower, which open from January onwards, are of normal shape and size, enclosing a tight rosette made up of a quantity of petals which are white with emerald green markings. The inner rosette cannot close but the complexity of the petals prevents insects from visiting and pollinating the ovary – as a result, there is no fertile seed. To compensate, a huge number of small bulbs is produced, pushed up through the soil and then carried down into woodland leaf mould to re-root.

Snowdrops flourish in northern Europe and in the northern states of the USA, but in the southern states, although they will flower, the scarcity of cold winter days makes them apt to dwindle from year to year.

All snowdrops prefer a fairly solid, heavy soil, moist but not too wet. Dividing them is one of the most satisfactory jobs – the golden rule is to separate them after flowering and before the

foliage starts to change colour. We try to remember to do a few every March, or even February.
Bulb   Zone (4-)

Galanthus 'Samuel Arnott' is dedicated to a former provost of Dumfries. The renowned horticulturist Henry Elwes subsequently allowed it to naturalize itself at Colesbourne Park, his home in Gloucestershire.

Long in bloom, strong in fragrance, the flowers have a deep green, heart-shaped mark round the sinus and are full and rounded, with stout stems and good leaves. These snowdrops will increase rapidly.
Bulb   Zones 2-8

Garrya elliptica, the silk tassel bush and a native of California, commemorates Nicholas Garry, secretary of the Hudson Bay Company. It was introduced into Britain in 1828.

In January and February it is particularly attractive for its silky catkins – silvery-grey and up to 30cm (12in) in length. The leaves are attractive too – rounded and shiny on top and woolly on their undersides.

Trimmed as a wall shrub, it makes a striking specimen, 3m (10ft) high and equivalent in spread. It can be used free-standing, but its leaves tend to become burnt by cold east winds unless it is planted in a protected position.

At the Regional Parks Botanic Garden near Berkeley, California, there is a cultivar, 'James Roof'. Raised as a selected seedling, it has extra long catkins, 35cm (14in) or more. Only the male form has these lovely catkins, the female's fruit being purple-brown.
Evergreen shrub   Zone 8-

Gaultheria procumbens, the checker- or partridge-berry, is American in origin. It was introduced into Britain by David Douglas in 1826. Highly valued in its native land, it was to inspire the first use of the term 'groundcover'. Clyde Bailey wrote in *The Cyclopedia of American Horticulture* in 1900 that 'Gaultheria is apt to form handsome evergreen ground cover.'

Along with the black-fruited *G. shallon*, this, the most reliable of the gaultherias, will make a dense thicket with arching branches up to 90cm (30in) high. It spreads by means of underground stems, which is why it makes such good groundcover. It has bright red autumn fruit, small, dark, glossy leaves that turn bronze and red in winter, and pink flowers in summer.
Evergreen shrub   Zones 3a-7a

Hamamelis 'Pallida' is a native of western and central China, described by Augustine Henry in 1888. This particular species was raised in the garden of the Royal Horticultural Society from seeds which were probably taken from the Kalmthout Nursery – long before it became the famous de Belder's Arboretum.

Reaching 2.5m (8ft) in height and with an equal spread, its strongly-scented flowers appear in profusion, sometimes as early as the end of December but with the flush of flowers appearing in January and February. It can be spectacular on a sunny February day, underplanted with winter aconites, species crocus and earliest narcissus.

Witch hazels, as they are commonly called, prefer well-drained, neutral-to-acid soil, so they will grow on lime but not on chalk. No pruning is necessary, but a judicious cutting of stems for the house does no harm.
Deciduous shrub   Zones 5b-9a

Hebe rakaiensis, one of the shrubby veronicas, owes its name to Greek mythology: Hebe, wife of Hercules, was the goddess of youth and a cup-bearer to the gods. Like most hebes, it comes from New Zealand.

Its attractive shape and pale green, glossy leaves make it an especially useful evergreen groundcover plant for the garden in winter. It forms a dense, compact mound 60 to 75cm (24 to 30in) high and slightly more across.

As neat as box but faster-growing and requiring no clipping, it can be used in a similar manner. However, because it is short-lived, I would not recommend it for topiary – a specimen will look well-groomed for 10 or 12 years and will then open out, losing its neatness. You can prune it hard back, but it will take 2 or

Garrya elliptica

Gaultheria procumbens

Hedera helix 'Glacier'

3 years to recover.

A good seaside shrub, it roots easily from cuttings, and with its abundance of fibrous roots may be transplanted with ease.
Evergreen shrub   Zone 7

Hedera colchica 'Dentata Variegata' is the cream and golden variegated form of the Persian ivy from the Caucasus and Asia Minor.

Its large leaves have a leathery texture and, as the name suggests, are toothed, with irregular markings which change from creamy-yellow to creamy-white as each leaf matures. Leaves concealed from the light can be entirely yellow.

This is a dramatic ivy when allowed to grow in great drifts along walls and fences – it will reach 9m (30ft). It becomes very heavy in a mass, in fact too heavy to be supported by its own aerial roots, and we find we must keep ours tied back securely.

The variety *H.c.* 'Dentata' has leaves which are even larger and entirely green; *H.c.* 'Paddy's Pride', correctly known as 'Sulphur Heart', has bolder, yellow markings.
Evergreen climber   Zones (8a-9a)

Hedera helix 'Glacier' is a useful and decorative form of the common ivy.

An attractive cool grey colour, it does particularly well on alkaline soil. It makes a lovely background for mauve, winter-flowering

Hedera helix 'Goldheart'

rhododendrons, and I like to use it in conjunction with snowdrops and early *Iris reticulata.*

Most ivies are native to northern Europe, evergreen and hardy, but their leaves may be scorched by hard frost and icy winds. At the end of every winter we 'groom' our plants, clipping off any growth which threatens to become invasive and giving them a good brush-down to rid them of any browning and withered leaves.
Evergreen climber   Zone 6b

Hedera helix 'Goldheart' must be one of the most popular cultivars of ivy. Its leaves are small, neat, three-pointed and green with a golden-yellow centre; the stems are inclined to pink or red.

A tidy, compact climber, it reaches 2.75m (9ft) and so is good for small gardens. It is best grown in sun to help keep its golden markings, although they will remain even in almost complete shade. It will grow as groundcover but looks best of all climbing through a tree which branches from near ground level. With us it is clambering through a quince and many of the longish shoots are now trailing downwards. Recently it has taken to flowering, and the berries are an attractive bright yellow in winter.

Another variety of the common ivy, the tough, handsome, reliable *H.h.* 'Baltica', makes ideal groundcover for growing under deciduous trees in east-coast American gardens where grass would succumb to summer heat.
Evergreen climber   Zone 6b

Helleborus corsicus is a native of Corsica, Sardinia and the Balearic Islands. Its generic name is thought to mean 'food to kill' – the drug made from its roots has been used both as cure and poison.

It carries clusters of pale apple-green flowers from December to April; its leaves are light grey-green with darker veins and scalloped edges. During spring and summer it throws up a mass of new 60 to 90cm (24 to 36in) shoots. A well-established plant may produce as many as 30 stiff stems in the right conditions – a cool, rich soil in light shade and shelter from strong winds and the worst frosts.

You should leave this plant undisturbed, give it a rich feed in autumn, and then watch out for seedlings around it, which you may dig up and grow on to make new clumps, treating the roots with benomyl to prevent die-back and botrytis. When the seeds have ripened, cut back the old shoots to ground level to make way for the new growth.

Lilies are among its best companions, their

young shoots enjoying the shade created by the hellebore leaves.
Evergreen herbaceous perennial   Zone 7-

Helleborus foetidus grows wild in southern Europe, western Asia and Britain. The description *foetid* is rather unkind, as it only shows this stinking characteristic when it is bruised or brushed against.

It is a striking rather than a beautiful plant, 60cm (24in) tall. Its handsome, deep green, fan-shaped leaves contrast well with the clusters of flowers. These, cup-shaped and greenish-yellow, appear almost fluorescent, their sepals finely margined with crimson.

It should be given the same treatment as *H. corsicus*, and can be grown in any shady corner, or under deciduous trees – Gertrude Jekyll underplanted her nut walk with hellebores of all sorts. It associates well with primroses, violets and snowdrops in woodland; for a more sophisticated effect, try it with grey foliage plants.

There is a handsome red-stemmed form, *H.f.* 'Wester Flisk', and a sweet-scented strain, *H.f.* 'Gertrude Jekyll'. Each should be kept separate if seedlings are to be true to type.
Hardy herbaceous perennial   Zone 6b

Helleborus niger, the Christmas rose, is a woodland and mountain plant native to southern Europe and parts of western Asia. In England it usually starts to flower in December, but in the USA. February is a more likely time.

The blooms, with golden stamens, start as pure white and then tinge to pink as the sepals age. The leaves are broad and leathery, divided into 7 or 9 segments, and low on the ground, so that the flowers, 10 to 12cm (4 to 5in) tall, stand above the foliage.

This hellebore is not easy to establish, so plant it carefully in half shade and in soil which does not dry out. Mulch with leaf mould each year and nurture well.

A fine form, *H.n.* 'Louis Cobbett', has large, attractive flowers flushed with pink. I also like *H.n.* 'Potter's Wheel' – when propagated by division, not grown from seed – and an exceptionally good hybrid, *H.* Nigriliv, with good dark leaves and clear white flowers.
Evergreen herbaceous perennial   Zones 4-8

Helleborus orientalis, the Lenten rose, is the most ancient of the hellebores.

The flowers are intensely beautiful, ranging from pure white through pink to wine-coloured. They are easy to establish and long-lived; luscious new leaves appear before the old ones have died down so there is never a bare patch on the ground where they have been planted. They will reach some 60cm (24in).

Increase them by seed, either allowing them to sow themselves or gathering the seed as soon as it ripens and putting it immediately in a seedbed or tray. Treat the seedlings and even the grown plants with benomyl to prevent the green leaves from browning.
Evergreen herbaceous perennial   Zones 4-8

Hepatica x media 'Ballardii', commonly known as the liverwort, derives its generic name from the Greek word for liver, *hepar,* and is traditionally thought of as a liver cure. This

Hepatica x *media* 'Ballardii'

Iris unguicularis

hybrid – a cross between *H. nobilis* and *H. transsilvanica* – was raised in 1916 by Ernest Ballard at his famous Colwall nursery near Malvern in Worcestershire.

You will love this beautiful plant with its clumps of intensely blue flowers, 10cm (4in) high, which appear towards the end of winter. The leaves follow the flowers and stay green throughout the rest of the year.

Closely related to the anemone, hepatica is to be found growing in half-shady woodland throughout the northern temperate zone, so plant it in a shady corner. It dislikes disturbance and should be divided in early autumn or even late summer when it appears virtually dormant. Seed can be sown in shallow drills in moist, preferably alkaline soil. Transplant the seedlings carefully in autumn when they are large enough to handle easily.
Evergreen herbaceous perennial   Zone 5

Heuchera 'Palace Purple' was introduced into Britain from the USA and is one of the best recent additions.

Of colour and interest throughout the year, the dark bronze-red leaves – light magenta and pink on their undersides – are heart-shaped and have irregularly cut edges. In May and June masses of dark, wiry stems carry feathery heads of tiny white flowers, which soon expand to rosy-bronze seed pods. Gather the seeds and sow them the following spring. Raising from seed gives a range of colours of which only the deepest should be called 'Palace Purple'.

This heuchera looks attractive with grey and silver foliage – santolinas, dianthus and artemisias – or contrasting with golden marjoram or the golden form of creeping Jenny (lysimachia).
Evergreen herbaceous perennial   Zones 4-9

Iberis sempervirens, commonly known as candytuft, comes as its generic name suggests from the Iberian peninsula, yet its common name is a reference to its supposed Canadian or Cretan origin.

The deep green, narrow leaves remain on this low-spreading sub-shrub all through the winter, and for many weeks in spring it is covered with the whitest of blooms.

Put it against a dark background where it will stand out, especially if planted in full sun. It is easy to raise from seed or from cuttings.
Evergreen perennial sub-shrub   Zone 4-

Idesia polycarpa is a fast-growing deciduous tree which will eventually reach 12m (40ft).

In summer the large leaves are distinctive

and the yellow-green flowers fragrant, but in winter its interest lies in the bunches of bright red berries, which hang on well until blackened by the frost. To have berries, you must plant both male and female, but not necessarily close together – at Kew the flowers are pollinated even though the trees are several hundred metres or yards apart. This, and the fact that they are scented, suggest that they are pollinated by insects attracted by their fragrance.

There are some trees and shrubs which flower and fruit better where the summer sun is hot enough to ripen the current year's growth thoroughly – I believe that idesia falls into this category. I have seen wonderful displays of fruit in the south of France, in the USA at Wavehill Botanic Garden in New York, and at the Morris Arboretum in Pennsylvania, where the bunches of berries were spectacular against a clear-blue winter sky.
Deciduous tree   Zones 6b-9a

Ilex x altaclerensis 'Golden King' is one of the countless cultivars bred from the common wild holly, *I. aquifolium.*

A small, rooted cutting of this gold variegated cultivar may take 18 years to reach a height of 2.75m (9ft). The edges of its leaves are smooth and banded with gold, the central area being dark green.

This holly produces lovely bold effects when clipped – hedges, balls and tiers, spirals even. For the imaginative topiarist the great joy of holly is that it grows direct from the bud to which it has been clipped – there is therefore little die-back.

Despite its name, this variety is female and will give you berries even when you have clipped it.
Evergreen shrub   Zones (7b-9a)

Ilex x altaclerensis 'Lawsoniana' is a cross between *I. aquifolium*, the English holly, and *I. perado*, the Azores or Madeira holly.

Its rich, dark green leaves have a central dash of gold surrounded by light green – a most unusual and attractive marking – This is one of my favourite hollies for the colour and lightness it gives to the border in winter. It will grow to 6m (20ft) eventually, but by judicious pruning can be kept to the height you require.

The stems are dark, as are those of *I. x a.* 'Hendersonii', of which it is a sport and to which it is inclined to revert unless you watch it carefully and cut out plain green shoots as soon as they appear.

A visit to the National Holly Collection at the

Valley Gardens in Windsor Great Park, Berkshire, will tell you all you wish to know about these indispensable winter trees and shrubs.
Evergreen tree or shrub   Zone 7b-9a

Iris danfordiae takes its name from the Greek rainbow goddess. A native of Turkey, it grows in the mountains up to altitudes of 2700m (9000ft), but mostly near the snow line.

The dainty, deep yellow flowers have falls that are pencilled with olive-green and crested with orange. The flowers open in February on 7.5cm (3in) stems.

After flowering, the bulbs have a habit of splitting up into little bulbs which take years to reach flowering size. To prevent this, dig up the bulbs in summer and plant them in a new site, 15 to 20cm (6 to 8in) deep in pockets of sand for perfect drainage. This greater depth of planting encourages them to 'concentrate' on making bigger side bulbs.

I have found that an application of bone meal in autumn helps this iris to flower regularly; it will also benefit from a good, thorough summer baking.
Bulb   Zones 4-9

Iris foetidissima – called variously Gladdon, Gladwin or the roast beef plant – is, rendered literally, the stinking iris. Its smell is not really unpleasant, more a reminder of Sunday lunch.

In winter the brilliant, coral-red seeds distinguish it from the brown or black seeds of all other irises. In summer it has small, dingy flowers of an inconspicuous purplish colour.

It prefers a shady, not too dry, corner, and requires nothing except admiration, especially on snow-covered winter days. Brought indoors, the seed heads remain in immaculate condition for weeks, but if you take Gertrude Jekyll's advice you will hang the bunch upside down for a period, to stiffen the stalks.
Bulb   Zones 5-9

***Iris histrioides* 'Major'** grows wild in central Turkey, south of the Black Sea.

Its large spode-blue flowers are outstanding for their beautiful navy blue veins and spots and distinctive orange blotch. The flowers appear before the leaves and are extremely hardy, undamaged by snow, frost or rain.

If planted in a sunny, well-drained position, they will increase by producing bulblets around the parent bulb, making an intense blue carpet, 10cm (4in) high, from January until late February.
Bulb   Zones 4-9

Iris reticulata grows wild in central Turkey, the Caucasus, central Iran and Iraq. It is found mostly on the snow line.

These irises with their 'netted' bulbs are essential for the garden in February, where they push their way through the snow to a height of 15cm (6in) and bloom with the earliest crocuses. The rich velvety purple flowers have gold markings on their falls and a faint but deliciously sweet scent which makes them a good candidate for being grown in pots in a frame for indoor use.

They do not, in my experience, increase rapidly, but the clumps will thicken, so a few of these inexpensive bulbs should be bought and planted every autumn. After flowering, the rush-like leaves become long and rather untidy-looking, but it is best to allow them to die down naturally.
Bulb   Zones 4-9

Iris unguicularis is more generally known as *I. stylosa* or the Algerian iris, native to Asia Minor, Algeria, Crete, Greece and Syria.

Its delicate violet-blue flowers which are 7.5cm (3in) across and 20cm (8in) high, change colour in every light. After a hot, dry summer, they will often appear before Christmas and flower right through to March if the weather stays mild.

Idesia polycarpa

Once established it should not be disturbed, but the best time to transplant it is when new roots are developing in August. Keep all browning leaves cleared away, lest they tempt snails and slugs.
Bulb   Zone 8

Itea ilicifolia is a native of western China, raised from seed sent to Britain by Augustine Henry in 1895.

Its chief glory in winter is its glossy, holly-like foliage, dark green on top and paler underneath. In summer it is covered with hanging racemes of fragrant, greenish-white flowers.

Of a looser habit than hollies or Portugal laurel, it will reach 4.5m (15ft) if frost does not damage it. It is slightly tender and likes the shelter of a wall. I grow it against a west-facing wall, where it has survived for 10 years, even though it has been cut back severely in hard winters. It will grow in sun or shade and likes a slightly acid and moist soil.

I. virginica 'Henry's Garnet', which grows as a low sprawling shrub, is unquestionably the best cultivar for the United States.
Evergreen shrub   Zone 9a-

Jasminum nudiflorum was introduced into England from west China in 1844 by Robert Fortune.

It is surprisingly versatile and may be treated as a climber, a free-standing shrub or as groundcover. It has no means of self-support, so its older stems must be secured to a wall or fence, and then the new growth will cascade down, making an elegant yellow waterfall from December right through to March. At the height of winter its yellow flowers are so welcome that you forgive it for being at other times rather an ordinary deciduous shrub. When grown along the ground, it should be combined with stiff-branched shrubs, such as *Cotoneaster horizontalis* or *C. dammeri*.

The only attention it needs is for some of the old wood to be cut out in April. You will achieve a natural effect if you cut each shoot back individually.
Deciduous shrub   Zones 6b-9a

Juniperus communis 'Hibernica', the Irish juniper, forms a wonderful upright, narrow, dense column of silvery-grey foliage, 3m (10ft) tall. Unquestionably one of my favourite evergreens, it is excellent as a formal specimen.

We have tied ours gently around with twine since one night in April several years ago. Heavy and unexpected snow fell, and we woke to find the branches of our lovely symmetrical

junipers bending wide open like water lilies, with a pool of snow in their centres. Careful husbandry restored them to shape.

You can clip them to the shape and size you require, but my experience has been that after 20 years or so they begin to lose their figures. This is the moment to be decisive: dig them out and replace them with carefully nurtured, younger specimens.

Tough, hardy and tolerant of chalk, they root from hardwood cuttings taken in late summer and prefer sun to full shade.
Evergreen tree   Zones 5b-8a

Kerria japonica takes its name from William Kerr, one of England's earliest plant hunters, who discovered it in China in 1803.

In winter it has shining canes that will be bright mid-green in colour if the older canes have been cut out after the plant has flowered in May. In early summer it is wreathed in rich orange-yellow flowers like oversized buttercups. *K.j.* 'Aureo-vittata' has clearly defined green and yellow bands up its branches.

Set behind the coloured stems of dogwoods or willows, kerria will become a useful backdrop, providing extra height (1.8 m or 6 ft) and depth. It also looks good in front of a wall.
Deciduous shrub   Zones 5b-9a

Lamium maculatum 'Beacon Silver' is a cultivar of the dead nettle.

It is hardy and most useful as perennial groundcover; its shining silver leaves, edged with green, flourish equally in light shade or full sun. As the leaves age mildew mottles them with reddish spots, which I find complement the deep mauve-pink May flowers.

It is a fast-growing carpeter 20 to 30cm (8 to 12in) tall, and makes ideal groundcover for early bulbs. A variety, *L.m.* 'White Nancy', is now available with white flowers and even greyer leaves.

Clip each plant down to ground level after it has flowered towards the end of June, and give it a top dressing of peat, compost or general fertilizer. Within days, new leaves will be pushing through; 2 or 3 weeks later an impressive new carpet will be forming again. To increase it, take cuttings or divide existing clumps.
Perennial groundcover   Zones 4-9

Leucojum vernum, the spring snowflake, is a member of the amaryllis family, native to Europe and Asia Minor.

In the northern hemisphere, its flowers – 15cm (6in) tall but reaching up to twice that in

Lamium maculatum 'Beacon Silver'

warm weather – appear in February. White with a green tip, they are shaped like Victorian lampshades.

Plant the bulbs 10cm (4in) deep in the autumn and leave them undisturbed until the clumps become obviously too thick, then divide them as the leaves yellow. I prefer them planted in drifts rather than in a block, perhaps with parallel drifts of narcissus and primroses. They like a moist soil which does not dry out in the hot sun. If you leave them to die down naturally they will increase readily.
Bulb   Zones 5-9

Ligustrum ovalifolium 'Aureum' is the semi-evergreen, oval-leaved golden privet from Japan.

Its rich yellow leaves, sometimes with a green centre, glow in the sunshine and make an outstanding splash of colour. The creamy flowers, which appear in August, are surprisingly sweet-smelling.

With a height and spread of some 3.6 to 4.5m (12 to 15ft), it grows well on any reasonable soil, is tough, and will often retain its leaves all winter through. If it has a fault, it is that it grows too quickly and needs cutting back.

I use it in the mixed border to create solidity and also as a host plant for early-flowering *Clematis macropetala*.
Evergreen shrub   Zones 7a-9a

Liriope muscari is a native of China and Japan. Like other lilyturfs of the *Ophiopogon* and *Liriope* species, it provides impenetrable groundcover and is grown extensively in the warmer parts of the USA.

It is a striking low evergreen plant with

Leucojum vernum

Liriope muscari

Lonicera x purpusii

narrow, dark green, strap-shaped leaves about 38cm (15in) long. Through these, in late autumn, appear violet stems bearing clusters of tight violet-purple bells.

It likes a well-drained soil, either in sun or shade, and when established will stand up to very dry summer conditions. It can be used as an edging along a shady pathway. Clumps are increased by division and replanting.
Perennial evergreen groundcover   Zones 6b-9

Lonicera fragrantissima, a shrubby evergreen honeysuckle, was introduced by Robert Fortune from China in 1845.

It will reach 1.8m (6ft) with an equal spread, and like the other winter-flowering honeysuckles such as *L. x purpusii* is definitely a bush, not a climber. From December onwards it is covered in small, creamy flowers that give out a clear, strong, pure scent. I rate it above the famed winter sweet for the intensity of its scent.

In summer it is one of those shapeless shrubs that needs enlivening – you might grow clematis through its branches – but one deep sniff of the flowers in winter will blot out memories of the plant's summer inadequacies. It has the rare quality of flowering best in a sunless position.

L. standishii and the hybrid of the two species, *L. x purpusii*, are similarly sweetly scented and winter flowering.
Semi-evergreen shrub   Zones 6-9a

Lonicera nitida 'Baggesen's Gold' is the offspring of *L. nitida*, introduced by E.H. Wilson from China in 1908; the golden variety was raised later as a seedling by the firm J.H. Baggesen of Pembury in Kent.

This golden shrub with tiny, glossy, evergreen leaves will reach 1.8m (6ft) and more. It likes a well-drained soil and must be planted in full sun if it is not to lose its golden quality. It is easy to propagate from young or hardwood cuttings, and is a wonderful standby for the flower arranger needing long, arching sprays in winter or small golden shoots in summer. The yellow-green flowers, which open in April and May, are insignificant.
Evergreen shrub   Zones (7-9a)

Lychnis coronaria alba, the white rose campion, is a native of northern Europe.

Its silver-grey basal leaves form rosettes which look lovely in winter; in summer, its flowers are of a milk-white colour.

These plants will reach 60cm (24in) if grown in full sun on a light soil, but they are short-lived. I plant crocuses or other early bulbs between them. Self-sown seedlings will spring up around the parent plants and these may be thinned and replanted to make a bigger group.

L.c. oculata is a white form with a cerise eye, which fortunately comes true from seed. There is also the species *L. flos-jovis*, which is pale cerise.
Herbaceous perennial

Magnolia grandiflora was introduced into England from the southern states of the USA in the 18th century.

Its large, glossy, leathery leaves, sometimes with rust-coloured undersides, make it an outstanding tree for both winter and summer effect. In winter the cone-like fruits with orange to red seeds are an added attraction. The scented, creamy-white flowers open from July to

September.

When raised from seed, this magnolia is unlikely to flower for 15 years, but from cuttings it may take only 7 years. In England it is best grown against a wall, where it will have some degree of protection. Pruning should be kept to the minimum; where it is grown free-standing no pruning is needed, as it forms a natural pyramid shape.
Evergreen tree or shrub   Zones 7b-10

Magnolia heptapetala (formerly *M. denudata*), the Yulan magnolia, was introduced into England from China under the sponsorship of Sir Joseph Banks. It has been grown in Britain for 200 years. William Frederick, the American landscape architect, believes that it should be 'a prime candidate for the garden which could have only one tree'.

With brilliant white flowers that open in mild weather in February, it is one of the most outstanding of all magnolias. Scales with shaggy grey hairs protect the buds until the moment when they begin to burst open, revealing petals that are thick, fleshy and highly scented. One prays for freedom from frost so that their beauty will not be spoiled as they open.

The tree will grow to 1.8m (6ft) three years after it has been planted; another five years will see it double in size.
Deciduous tree   Zones 6a-9a

Magnolia kobus stellata, the star magnolia, was introduced into England from Japan in 1862.

The earliest of the smaller white-flowered magnolias, this beautiful shrub has starry, fragrant flowers. The furry buds are so

Magnolia x *soulangiana* 'Lennei'

attractive and the bark so aromatic that it does not disappoint even if, in a normal winter, it delays flowering until March or even April.

It will reach 3m (10ft) in height and spread to 4.5m (15ft), so you must allow it enough space. Plant early spring bulbs – scillas and chionodoxas – under its low-spreading branches to flower before the magnolia buds open.
Deciduous shrub   *Zones (5a-9a)*

Magnolia x soulangiana 'Lennei', a hybrid of *M. heptapetala*, is named after Herr Lenne, gardener to the King of Prussia. Its parent was raised by M. Soulange-Bodin at Fromont near Paris, but this hybrid is said to have been found in a garden in Lombardy in the mid-19th century.

It qualifies for the winter garden for the beauty of its vivid rose-purple buds, which gradually display white as they open up into large, pink, cup-shaped flowers. Unless the buds are caught by a late frost at the end of April, they seldom suffer damage. Its long leaves are light to grey-green.

Vigorous and with a spreading habit, you should allow it plenty of space, particularly since it looks best as a specimen in full flower in April. It is one of the best magnolias for a garden on alkaline soil.
Deciduous tree   *Zones 5b-9a*

Mahonia aquifolium, the Oregon grape, is a native of western North America, introduced into England in 1823.

Its glossy pinnate leaves turn red at the first touch of frost in winter and the first flowers appear in February, to be followed by clusters of fragrant yellow flowers in April and May. These are then followed by an abundant crop of black berries with a purplish bloom.

One of the hardy broad-leaved evergreen shrubs, it is the hardiest of all the mahonias and has proved a survivor even in Illinois, in the Morton Arboretum. Easily propagated by seed, it will also root from cuttings, and is not fussy about soil. It will grow to 76cm (30in) under the right conditions.

Unfortunately, most nurseries sell seed-raised plants of uneven height and quality, inferior to selected, named clones such as *M.a.* 'Apollo'.
Evergreen shrub   *Zones 5b-9a*

Mahonia x media 'Charity' commemorates an act of charity – a gift from the nurseryman L.R. Russell to Sir Eric Savill of a seed of *M. lomariifolia* from the Slieve Donard nursery in Northern Ireland. When it grew, it turned out to have *M. japonica* as its other parent.

Its soft, primrose-yellow flower spikes, 23cm (9in) long, may well first appear in November, but the buds will go on opening for weeks. The leaves are broad and very prickly, so do not plant it where you may brush against it. Unpruned, it will grow to 2.5m (8ft), but if long shoots are pruned back after flowering, the bare stems will break into lateral growth.

It does well in a shady corner and is lovely to come upon in a woodland garden. With its lily-of-the-valley scent, winter flowers and architectural shape, it is also ideal for a town courtyard garden or as a feature near the front door. It is very hardy and makes a good host for late-flowering clematis.

There are a number of other good named forms of *M.* x *media*; 'Lionel Fortescue' is one of the best.
Evergreen shrub   *Zone 9*

Malus 'Red Jade' – known in the USA as the Red Jade crab-apple – was raised at the Brooklyn Botanic Garden in 1935. The crab-apples as a race are beautiful both in fruit and flower; they are among the best of the small trees for their winter fruiting.

The small, bright, cherry-red fruits of *M.* 'Red Jade' last well into winter. They show up even before the leaves fall, since they hang on 5 to 7.5cm (2 to 3in) stems, lower than the short-spurred leaves. The weight of the fruit makes the naturally hanging branches even more pronouncedly pendulous. As if that were not enough, attractive pink buds open in May to reveal white flowers.

I think it is one of the best small trees of recent introduction. It makes an attractive specimen on a lawn, and its leafless canopy is an ideal host for the earliest crocuses and snowdrops. The rather haphazard way its branches grow give its canopy extra charm.
Deciduous tree   *Zones 4-8*

Malus x robusta, incorrectly known as the red Siberian crab, was first cultivated in Britain about 1815. Its round, cherry-sized, bright scarlet fruits are decorative throughout the winter and, together with its yellow-fruited cousin, *M.* 'Golden Hornet', it carries them for longer – from late autumn until February most years – than the other crabs. Only the hardest of frosts will finally cause them to fall.

I prefer these crab apple trees planted in a group of three. Their moments of glory – while they blossom and when their fruit turns colour – are important but brief, so I like to keep them as background trees. I might feel differently if I lived in a town and had a very small garden.
Deciduous tree   *Zones 4b-8a*

Myrtus luma, recently renamed *Luma apiculata* and known in the USA as *Amomyrtus luma*, is a myrtle which grows naturally in Chile.

It is the bark of this small tree which is all-important in winter. Cinnamon-coloured, it peels off in patches to reveal a beautiful cream-coloured inner layer. Like all the myrtles, it is an aromatic evergreen which produces a generous crop of solitary white flowers in late summer and early autumn. The flowers are followed by red and black fruit, which last well into winter and are edible.

A tree for warm, temperate and tropical regions, it is best grown against a south-facing wall. It can grow up to 7.5m (25ft) high and is surprisingly good for coastal planting. It likes a well-drained and rich soil, but is just about lime-tolerant.
Evergreen tree   *Zone 9*

Narcissus bulbocodium, the hoop petticoat daffodil, takes its generic name from the Greek youth so obsessed by his own reflection that he was turned into a flower by the gods.

This dandelion-yellow miniature narcissus is very pretty. Each year at Wisley in Surrey, they provide a beautiful golden expanse across the alpine meadow in February or early March. Nearby, the Savill and Valley Gardens in Windsor Great Park both enjoy similar fields of the cloth of gold.

A colony planted in rough grass will spread of its own accord, provided the soil is fairly moist. Their 30cm (12in) rush-like leaves have a lengthy growing season, starting to show themselves quite early in the autumn. If planted in cultivated soil, they tend towards congestion unless planted a good 10cm (4in) down.
Bulb   *Zones 6-9*

Narcissus 'February Gold' is a hybrid of *N. cyclamineus*, and its apt special name tells us all we need to know – that it is golden and that it flowers (although not always) in February.

It is a must for the garden in winter: it is a good dark yellow, quite tall – 30cm (12in) high – and, unlike most of the Cyclamineus varieties, has a relatively short, trumpet-shaped cup. Its petals sweep up and then back.

Ideal for the front of the border, I grow these early-flowering narcissi with *N.* 'Tête-à-Tête' and scillas, which are fine companions. Wherever possible, I grow them between and through low herbaceous plants: I have planted some around the base of *Clematis* x *jouiniana* whose stems effectively cover the dying bulb leaves as it puts on its spring growth.
Bulb   *Zones 5-9*

Osmanthus heterophyllus has a botanical name which is a tribute to its sweet-smelling flowers – *osme* is the Greek word for fragrance, *anthos* for a flower.

A large bush of this important member of the club of late-winter-flowering, evergreen shrubs is both an impressive sight and a fragrant experience. It flowers prolifically and its scent is strong as well as sweet. The small ivory, tubular blooms push themselves out from the rather stiff twigs and dark green foliage.

A neat, reasonably hardy shrub, it will grow to about 3m (10ft) given time and will thrive in sun or shade, acid or lime.
Evergreen shrub   *Zones (8b-9a)*

Osmaronia cerasiformis, now known as *Oemleria cerasiformis*, was introduced into England from California in 1848. Its common name is the Indian plum or oso berry.

In February and March drooping racemes of fragrant white flowers with green calyces appear, their distinctive almond scent wafting mysteriously on the air. When cuttings are brought indoors, the fragrance is overwhelming.

It is a useful but not showy deciduous shrub, with a habit similar to the blackcurrant – even its flowers are ribes-like. Its stems spring from the ground in great numbers and eventually form a great thicket, 1.8m (6ft) high and wide. Its leaves are green and glabrous above, downy and greyish underneath, and the plum-like fruit, 1cm ($^{1}/_{2}$ in) long, are brown when they first appear, purple when they ripen.

Both hardy and trouble-free, it is easily increased by rooted pieces or from seed.
Deciduous shrub   *Zone 4-*

Pachyphragma macrophylla takes its rather obscure-sounding generic name from the wall that divides its seeds – *pachys* is the Greek word for thick, *phragma* for screen or partition.

An excellent groundcoverer, the veins and stems of its round, distinctive leaves become tinged with purple in winter. Early in March, multitudes of bright white, starry, cress-like flowers appear on the leafy plants.

Mahonia x *media* 'Charity'

Narcissus 'February Gold'

Pachyphragma macrophylla

It likes the same sort of position, under shrubs and trees, as brunnera and symphytum. I allow the old leaves to remain until late February, when the new leaves and flowers begin to push through.
Evergreen groundcover   Zone 7

Pachysandra terminalis, Japanese in origin, is a rich green groundcoverer, a choice for the *cognoscenti*.

Best in winter, it looks particularly good planted as a carpet at the base of trees with interestingly coloured bark. Its glossy evergreen leaves grow in rosettes and are slightly toothed. It flowers, though insignificantly, in April, with terminal clusters of greenish-white blooms tinged with purple.

It is a slow starter but, once established, can easily be increased from summer cuttings or by division. It does best in a lime-free soil.
Evergreen groundcover   Zone 5

Parrotia persica must be a very hardy tree, for it comes from the mountain slopes south and west of the Caspian Sea.

Being a member of the witch hazel family, it blooms very early in the year. The flowers, small though they are, look especially lovely when snow is on the ground, consisting of stamens stained grape-juice crimson, hanging from seal-brown bracts. Just as they are about to open, pick some short branches to bring indoors.

It is a spreading tree, with branches going out at 45° from its trunk, creating a wonderful mushroom shape some 5m (18ft) high and 4m (15ft) wide. A few branches grow more vigorously than the rest, giving the mushroom an irregular outline. Its habit of growth attracts me as much as its leaf colour, although at certain times of the year its striking colours stop you in your tracks. I first saw this tree at Westonbirt in the autumn, looking resplendent with crimson and gold foliage.
Deciduous tree   Zones 6a-9a

Pernettya mucronata is a native of Chile and the Argentine.

The winter charm of this low shrub lies in its clusters of berries, ranging from white to deep pink. Provided a known male clone is planted, a good crop of berries will appear on the reddish stems of the females, lasting well into the winter. The berries of *P.m.* 'Sea Shell' start a soft pink and deepen in colour as the weather gets colder.

The dense yet lax habit of growth and the long, glossy, tooth-edged leaves make this family of low shrubs ideal for planting in large drifts. A group of them will soon form a thicket, 90cm (36in) high, which in summer will be covered with white, bell-like flowers. They do best on acid soil.
Evergreen shrub   Zones (7b-9a)

Phillyrea latifolia is a native of the Mediterranean region and has been cultivated in Britain since the 16th century. I am surprised how few visitors to my garden know about it.

It is a shapely shrub with broad, glossy, deep green leaves, and scented, ivory-white flowers in spring. It clips well, but left to grow naturally will make an elegant small tree, some 1.8m (6ft) high, after ten years.

It does best in light, open soil and full sun and will even transplant. We moved a large specimen, 2.5m (8ft) high and 1.5m (5ft) wide, without causing it any set-back, but we find it difficult to propagate from cuttings.
Evergreen shrub   Zone 7-

Phormium tenax, the New Zealand flax, was first introduced into Britain in 1789, but is to be seen naturalized in the west of Ireland.

Parrotia persica

Phormium tenax

Flowering in late summer, mature clumps, some 3.6m (12ft) high, can be relied upon to produce inflorescences of dusky red, tubular flowers in abundance. These are succeeded by black seed pods that appear in clusters all along the stem.

They grow best in moist, rich, loamy soil and look best in full sun. Ideal by water, they also make good framing plants for paths. Variegated and purple-leaved forms (many of them hybrids with *P. cookianum*) are perhaps even more striking, but are also less hardy.

Evergreen perennial   Zone 8

Picea pungens glauca 'Koster', a striking silver-blue form of the Colorado spruce, was first cultivated in 1865.

It grows into a narrow 1.8m (6ft) cone, with branchlets in horizontal layers. A small tree it may be, but it stands out in any setting and so needs careful placing in the garden if it is not to become too dominant.

The foliage shows up well against mellow red brick, and in my Cotswold stone country-side I prefer these bluer conifers in winter light. When frost is on them, the leaves look icy-blue; sprinkled with snow they are a gorgeous sight.

Picea pungens glauca 'Koster'

Two other forms of *P. pungens,* grown by Blooms of Bressingham in Norfolk, are *P.p.* 'Globosa', which eventually reaches 60cm (24in) and has a bush habit and a flat top, and the slightly taller *P.p.* 'Hoopsii', which has leaves of a light silvery blue.

Evergreen tree   Zones (3a-7b)

Pieris japonica came into cultivation in England in 1870 from a seedling raised by the nurseryman K. Wada of Japan.

An evergreen shrub about 1.2m (4ft) high with a bushy habit, its leaves are dark green and glossy. In winter, its young foliage takes on attractive red and bronze tints, and few sights are more delightful in February than the snow-white tassels of its bell-like flowers. The bells hang on after the pink buds have opened, finally browning over only in April.

There is a variety called 'Christmas Cheer' which flowers towards the end of December, if not necessarily on Christmas Day itself. Its flowers are pink with dark carmine-rose shading at their tips, giving a bi-coloured effect.

As a race, pieris love damp acid woodland soil, and preferably a sheltered position with some western sun.

Evergreen shrub   Zones 6a-8a

Pinus bungeana

Pinus bungeana, the lace-bark pine, was introduced from China in about 1846 by Robert Fortune.

This pine has exceptionally striking peeling bark, which is always an important feature in the garden in winter. In their natural setting in the Chinese mountains, some of the older trees have almost white trunks, but in the West we have to settle for dark apple green and yellowish patches showing through the warm brown bark.

Often multi-stemmed with its main branches growing vertically, it will reach about 18m (60ft), with a spread of 13.5m (45ft). Its needles are grouped in bunches of 3, with a tinge of yellow to their green; when you bruise them you will find them aromatic.

This pine has something to offer all the year round. If you want to see a fine specimen in England, go to Kew or Wisley in Surrey; in America, visit the National Arboretum in Washington DC, the Arnold Arboretum in Boston or the Morris Arboretum in Pennsylvania.

Evergreen tree   Zones 5a-8

Pittosporum tenuifolium takes its generic name from its resinous-coated seeds – *pitta* is the Greek word for pitch, *spora* for seed. It was imported into England some 100 years ago from New Zealand.

A hardy evergreen with an upright, pyramidal shape, it has shiny, olive-green leaves which are set off to advantage by its very dark stems. In summer it flowers with tiny, almost chocolate-brown bells; in warm areas they can be quite honey-scented if they are growing on a mature shrub.

It can be grown either as a specimen or as a wall shrub. A steady if unspectacular grower, it will reach some 1.5m (5ft) after 5 years, adding a foot a year until it achieves 6m (20ft). Like others of its kind, it enjoys salt sea breezes in more temperate regions.

Evergreen shrub   Zone 9

Platanus orientalis has been cultivated in England since the 16th century. With *P. occidentalis* – the so-called buttonwood of the eastern half of North America – it shares the parentage of the well known and much loved London plane, *P.* x *acerifolia*.

This, the oriental plane, is a most stately tree, 6 to 9m (20 to 30ft) high. Summer-leafing, it is very picturesque with its low, branching habit and the characteristic pale patches on its trunk.

You will see it in many old-established gardens, but I remember a particular specimen

Polystichum setiferum divisilobum

Primula denticulata

at Blickling Hall in Norfolk, where the lower branches sweep down and have layered themselves. It is underplanted there with *Lilium pyrenaicum*, now naturalized in the grass.
Deciduous tree   Zones 6a-9a

Polystichum setiferum divisilobum, commonly known as the soft shield fern, has a generic name describing exactly the way the sori, the divisions of the fern leaf, grow along the fronds – *polys* being the Greek word for many, *stichos* for a row.

This elegant cultivar has beautiful foliage. Immense fronds, 60cm (24in) long, spring obliquely from a soft brown crown, like green feathers stuck decoratively into an 'oasis' by a flower arranger. The spine, or rachis, of each frond swerves away from the crown, making the frond look as if it is searching out its own place in the sun. In fact, they do best in dappled shade and are very good under deep-rooted trees, such as oaks or evergreens.

Although slow to grow, they can be propagated easily by placing a leaf on peaty compost and then pegging it down.
Hardy evergreen fern   Zones 5-9

Populus alba, the white poplar, is a decorative tree all through the year. In winter I like the fine tracery of its branchlets. It has 3

outstanding virtues: it will grow in wet ground; it will grow rapidly; and it will cost nothing if you heel in a twig. My advice is to plant a specimen behind other trees. It will soon outstrip them and provide you with a tall background tree.
Deciduous tree   Zones 4a-9a

Primula denticulata, the drumstick primula, is a native of the Himalayas.

It has pale purple flowers with yellow eyes which appear on mopheads,30cm (12in) high, almost at the end of winter. Excellent for the front of a border or in a rock garden, it will also look effective in a wild garden.

It likes fairly rich soil, preferably one that stays moist and will increase naturally by producing shoots round the parent plant – with care these can be pulled off to make new plants. Good forms can also be propagated easily from root cuttings.
Herbaceous perennial   Zones 6a-8

Primula vulgaris, the common primrose (the English primrose in the USA), is a Linnaean name drawn loosely from its capacity to flower early in the year – *primus* being the Latin word for first. More significant is the name it was known by during the medieval period, *prima rosa*, meaning literally the 'first

rose' of the year.

Its flowers are a delicate soft yellow with a darker eye, all set off by its distinctive dark green leaves.

Primroses like dappled shade – I grow them under variegated holly, the yellows complementing each other. They grow best in moist areas and look attractive with snowdrops, early crocuses, scillas or violets.

Easily grown from seed, you should plant out all the well-grown seedlings in the autumn if you want them to be the 'first roses' of your new year.
Herbaceous perennial   Zone 6-

Prunus incisa 'Praecox', a form of the Fuji cherry, is a small tree raised by Hillier's nursery. I love to walk round the garden and find its blossom, knowing that it is merely the herald for a galaxy of beautiful flowering cherries soon to come.

Its numerous delicate flowers, white tinged with pale pink, open in February. In autumn the foliage turns scarlet, adding another virtue to this delightful shrub-cum-tree.

Most suitable for a small garden, it reaches a height of 6 to 10m (20 to 30ft). This prunus needs shelter so that the flowers do not get spoilt by the wind or the buds burned by frost. Normally planted as a specimen, it can be used to make an unusual and most attractive hedge.

If you decide to plant this cherry, make sure you get it from a nursery where it is grown out-of-doors, and plant it in October or early November so that it has time to settle before the onset of winter.
Deciduous shrub or small tree   Zone (6-)

***Prunus lusitanica* 'Variegata'** is an interesting version of that underestimated Victorian favourite, the Portugal laurel.

Its green and white foliage is lustrous, developing a pink flush in winter; its white,

hawthorn-scented June flowers, carried in long racemes, are sparkling. A splendid hedging plant, it is equally impressive planted as a specimen shrub or tree.
Evergreen shrub or tree   Zones (7b-9a)

***Prunus mume*,** the Japanese apricot, was introduced from Japan as long ago as 1844.

From the end of December onwards, richly fragrant, pale pink, single flowers appear on the bare dark stems of this small tree, which can be grown either as a specimen in a protected

Prunus mume 'O-moi-no-mama'

garden or trained up a warm south-facing wall.

If your garden is small, give this apricot a position at the back, in the corner, or at least 1.8m (6ft) away from the boundary fence or wall.

It is best to choose one of the more recently introduced cultivars, which will eventually grow 6m (20ft) high and 3m (10ft) wide – *P.m.* 'Benichidori', a soft crimson double, for example, or *P.m.* 'O-moi-no-mama', a semi-double white.
Deciduous tree   Zones 7a-9a

Prunus serrula (tibetica) was introduced into Europe from west China, first by E.H. Wilson in 1908 and then again by George Forrest in 1913.

One of the best winter sights, this cherry is outstanding for its wonderful mahogany-coloured bark, which peels away in horizontal bands leaving a strikingly smooth, polished surface.

You can see specimens at Westonbirt in Gloucestershire, at Bodnant in Wales and in other great gardens. Not a very large tree, it will reach 4 m (13 ft) after 5 years and 6.75 m (23 ft) after 13 years – its rate of growth slowing as the years pass.
Deciduous tree   Zones 6a-8a

***Prunus subhirtella* 'Autumnalis'**, the autumn cherry, is the most valuable winter-flowering cherry.

Its semi-double, white blossoms open on bare twigs from November through until March, and what they lack in size they make up for in abundance.

It can be grown either as a small, round-topped, slightly spreading tree, about 6 m (20 ft) high, or, with the central trunk removed, as a large twiggy shrub with a host of buds at eye level.

I first came upon it as an ornamental tree 10 years ago and stood back in amazement and wonder. I have still not planted one in my garden, and as each winter approaches I am regretful. Do not follow my example. Go out at once and find a nursery that supplies this admirable cherry.
Deciduous tree   Zones 5-8

Pulmonaria rubra takes its generic name – as its common name, the lungwort, suggests – from the Latin word *pulmo* for a lung, a reference to its supposed healing properties.

This form of the lungwort has long, narrow leaves and warm, soft red flowers which may open as early as January. Unlike other members of the family, its foliage is unspotted and a plain, fresh green, but its stalks are typically hairy.

An ideal spreading plant for groundcover, it loves to be cool and therefore in shade. It looks well with dwarf narcissi or under deciduous trees where to all intents and purposes it will be evergreen.

Pulmonarias, if they are to flourish, like to be divided from time to time – alternate years or every third year. If the soil is kept slightly moist, they will develop quickly into weed-smothering clumps. The plants will look tidier in summer if you spend a little time cutting off the flower stalks at ground level after flowering.
Herbaceous perennial   Zone 6

Pulmonaria saccharata 'Alba', the white form of the Bethlehem sage, is not to be confused with the common lungwort, or 'soldiers and sailors', which is markedly smaller.

Its snow-white, pearl-sized flowers emerge in mid-winter to be followed by large rosettes of clearly marked leaves which give interest for the rest of the season. Among the first plants to flower, regardless of weather, it is especially useful for a winter bunch.

Excellent as groundcover, it likes to grow in partial shade and will flourish best in moist soil. Plants can be divided in October or March – propagation by seed is usually unsuccessful.
Herbaceous perennial   Zones 4-8

Puschkinia scilloides, the striped squill, is dedicated to the 18th-century Russian plant collector, Count Apollos Mussin-Puschkin, whose hunting grounds were the Caucasus and Ararat.

Its bell-like flowers, pale blue or white, appear at the end of winter. The actual flower spikes are 10 to 12cm (4 to 5in) high – taller than most of the scillas and chionodoxas alongside which it grows in Turkey and Iran.

This squill prefers shade and looks good either in the wild garden or under deciduous shrubs or trees; it will grow in the sun but only if the soil is well-drained. The bulbs should be planted in bold drifts among shrubs in a border, so that they will show themselves off before the shrubs start to put out leaves.
Bulb   Zone 3-

Pyracantha atalantioides *(syn. P. gibbsii)* is commonly known as the fire thorn – *pyr* is indeed the Greek word for fire, *akantha* for a thorn. It was discovered in China by E.H. Wilson in 1907 on his first commission from the Arnold Arboretum in Boston.

Bright scarlet bunches of black-eyed berries cover this large, robust evergreen shrub from September until March – that is, if cold weather does not make the birds get to work too early. Its snowy flowers are far less interesting.

Frequently grown against a wall and clipped, this useful shrub looks just as good

planted as a single specimen. It will achieve a height and spread of some 3 to 4.5m (10 to 15ft). If you choose to have it on a wall, remember to prune it back hard after the berries have fallen if you want to keep it in good shape, but don't forget its thorns or they will give you a sharp reminder.
Evergreen shrub   Zone 7

Quercus ilex, the holm oak, is more generally and confusingly known as the ilex. Literally it is the holly oak – *quercus* being the Latin for oak and *ilex* for holly.

Its leaves are lustrous, soft olive-green on top and sage-silver beneath. Left to grow and with space to do so, it will make a vast round-headed tree. It will achieve 45cm (18in) of growth a year, but it is a slow starter and does not like to be moved. Although it does not like real cold, it will grow beside the sea.
Evergreen tree   Zones 8-9a

Rhamnus alaternus 'Argenteovariegata', the Italian buckthorn, takes its name from the Greek vernacular name for various spring shrubs. The green-leaved species of which it is a variety came originally from the Mediterranean.

It is a hardy evergreen shrub with silver-margined leaves, 3cm (1 1/4 in) long, which are shiny back and front. Its other great merit for the winter garden is its bright red fruit, which come in such profusion that they almost rival

Puschkinia scilloides

Rhododendron 'Nobleanum Aurum'

Ribes laurifolium

those of the holly.

Pyramidal in shape, it is a fast grower, reaching 3.6m (12ft), so if you plan to put it in your border you would do well to confine it to the back. It does not like to be moved, and as it is slightly tender, it fares better if given the protection of a wall.
Evergreen shrub   Zone 8

Rhododendron Nobleanum takes its generic name from the Greek vernacular description of an oleander – *rhodon* is a rose, *dendron* a tree. A hybrid – a cross between *R. arboreum* and *R. caucasicum* – it was raised by William Smith of Norbiton as long ago as 1829.

R. Nobleanum and its scions will, given the right conditions and the right season, be in flower by Christmas. The flowers can be white, pink or deep crimson, according to which form you choose, but this particular rhododendron has brilliant rose-scarlet buds in compact trusses, which open to rich rose and are flushed inside with white and spotted with crimson. Its leaves are a dullish, dark green and covered on the undersides with a rust-brown indumentum.

A large shrub or a small tree, it is one of the hardiest of its kind but a slow grower.
Evergreen shrub or tree   Zone 7-

Ribes laurifolium is the Chinese equivalent of the flowering currant.

Attractive, drooping sprays of sulphur-white flowers open on the male plants in February or early March. The less attractive flowers on the female plants are followed by red berries, which gradually turn blackish. For a good winter display, use three plants 45 cm (18 in) apart.

An excellent evergreen shrub, of dwarf size and with leathery, shiny leaves that are typical of most of the currants, it can be left to swirl over paths or stonework. In the border, it needs the protection of a wall.

Easily propagated from hardwood cuttings in winter or semi-ripe cuttings in summer, it needs little pruning.
Evergreen shrub   Zone 9

Rosa 'Ballerina' was first introduced in 1937. It is a vigorous, healthy bush rose that looks well planted as a low hedge, reaching 1.2m (4ft) tall and about 1m (3ft) wide, or it can be grown as a standard. All year it shows an abundance of glossy green leaves and from summer through to autumn carries masses of tiny flowers on hydrangea-like heads.

The blossoms are five petalled, pale pink edged with darker rose, and musk-perfumed.
Evergreen shrub   Zone 6

Rosa 'Ballerina'

Rosa banksiae, commonly known as the Banksian rose, honours Lady Banks, wife of the 18th-century naturalist Sir Joseph Banks who travelled with Captain Cook on the *Endeavour*. The rose was introduced to Europe in 1796. The double white form did not appear in Britain until some 11 years later, when it was sent to Kew by William Kerr – another inveterate plant hunter commemorated by *Kerria japonica*.

R.b. is a wonderful sight at the end of winter, covered with pendulous trusses of small, fragrant buds, which open as white flowers and continue through spring.

Given a sunny wall to ascend, it is vigorous and a great climber. I have seen it in the South of France as high as 12m (40ft). The double buttercup-yellow *R.b.* 'Lutea' is the most hardy –

it requires very little pruning.
Evergreen climber   Zones 8a-9a

Rosa 'Félicité Perpétue' is named after the daughters of the French rose grower A. A. Jacques, who first raised this hybrid from *Rosa sempervirens* in 1827.

A healthy rambler that flowers in profusion, its great virtue is that it is almost evergreen. It has light green, quite dainty leaves which hang on to the bitter end. If you grow it up an established tree it will increase with long new shoots each year, reaching an eventual height of 6m (20ft) or more. It needs to be pruned only lightly – taking the flowering shoots back and cutting out dead or old wood.
Semi-evergreen climber   Zone 6

Rosa 'Frensham' is a hybrid Floribunda – a seedling crossed with 'Crimson Glory'. It was raised by A. Norman in England in 1946.

Ruby velvet flowers in profusion are its summer joy; glossy, mid-green foliage its winter glory. It is an energetic grower – up to 1 or 1.2m (3 to 4ft) high – and has long been a favourite for large beds or hedges. Vita Sackville-West recommended it for a border largely composed of flowering shrubs. Ever colour-conscious, she conceived of it being planted in a purple or pink group alongside the other crimson Floribunda 'Dusky Maiden', musk roses such as 'Wilhelm', 'Pink Prosperity', 'Cornelia', 'Felicia', and the common old red herbaceous peonies, with Darwin tulips scattered among them.
Evergreen shrub   Zone 6

Rosa 'Poppy Flash' is a relatively new rose, introduced by the French rose-growers Meilland in 1971. It is a vigorous, bushy grower with stunningly beautiful blowzy double blossoms, 8cm (3in) across, in bright vermilion. The flowers form an open shape in 20-petalled clusters and with a delicious fruity fragrance. This floribunda rose blooms on into winter until the first frost falls on it.
Evergreen shrub   Zone 6

Rosa rugosa 'Rubra', the Ramanas rose, was introduced in 1845 by Phillip von Siebold, the Bavarian eye specialist turned voracious plant hunter and author of the classic *Flora Japonica*.

A hedge of this rose, which we planted several years ago as a windbreak, is now one of the joys of my garden the whole year through. A mass of decorative orange-scarlet hips, large, rounded and up to 2.5cm (1in) across, appear at the end of the flowering season and stay there – weather and birds permitting – through much of winter. The leaves darken in the autumn as the hips first appear.

I clip my hedge in winter to keep it 90cm (36in) high, which later on is the perfect height for smelling the aromatic, short-lived flowers.
Deciduous shrub   Zone 2

Rubus cockburnianus was introduced from China in 1907. *Rubus* is the Latin word for raspberry – a reference to the edible fruits of some of the species.

A vigorous grower, this bramble has dark purplish stems which are covered with a silvery-white bloom. It should be planted with a dark hedge as a background, preferably in the angle between two evergreens, to show off its stems in winter and its silvery-grey-backed foliage in summer. Its autumn fruits are deepest black.

I should warn that, with its 3.6m (12ft) arching stems, it is not a plant for the small garden. It is, however, a plant for the wild garden, where it can rampage without let or hindrance. It should be cut back hard in the spring.
Deciduous shrub   Zone 6-

Rubus phoenicolasius, the Japanese wineberry, was introduced from Asia in about 1876.

Remarkable for its brilliant red, hairy stems, this deciduous shrub sheds its leaves in time for its winter glory to be revealed. As if that were not enough, bright, orange-red, rounded raspberry fruits, which are both sweet and edible, appear from August onwards. Small wonder that the birds make this rubus a target. Its clusters of pale pink and white flowers appear in July but are unimportant compared with its striking winter stems.

A strong grower, reaching a height of 1.8 to 2.5m (6 to 8ft) and a spread of 1 to 1.5m (3 to 5ft), it makes a fine sight if planted alongside water, particularly in windy weather, when the under-surface is freely exposed. I have mine growing through railings near a pond – I like the contrast of the slate-blue railings against the brilliant red stems.
Deciduous shrub   Zone 6-

Salix alba 'vitellina', the golden willow, known in Europe since Roman times, is one of my favourites for its coloured stems.

The ideal place for it is by water, but in my garden I have 2 on poor, dry soil and they look wonderful in winter with their wands,90cm (36in) long, of brilliant egg-yolk yellow. Left unpruned, they will grow to a height of some 9m (30ft) and spread to about 4.5m (15ft).

All willows root readily; if you are willing to be patient, a stem placed so that at least half its length is below ground will save you buying a plant. To get the best effect you should coppice the stem every spring just before the sap starts to rise. I have chosen to do this about 30 cm (12 in) above ground level and now there is a gnarled head the size of a croquet ball from which an array of eye-catching golden stems radiate, creating in their density an impressive colour effect. *S.a.* 'Britzensis', the scarlet willow, may be treated in the same way.
Deciduous tree   Zones (3a-9a)

Salix daphnoides 'Aglaia' is the best form of the violet willow, and a look at its deep purple young shoots – overlaid with white bloom – will tell you how it has gained its name.

The joy of this tree comes from the large, pale purple catkins of the male plant, which develop before the leaves. On sunny days in February and even earlier, the distant effect of these against the blue sky is like a haze of flowers; when you get closer, you are sure to hear the bees at work.

Left unpruned, it will become a small tree, about 4.5m (15ft) high, but it will respond to coppicing, although I find that the best catkins come on the second-year wood.
Deciduous tree   Zone (5-)

Salix irrorata – literally, the dewy willow – is a native of the south-west of the United States, growing wild in the mountains of Colorado and Arizona. It was introduced into England in 1898.

It is the willow I enjoy most for its winter stems – vigorous young and green in autumn, they turn to purple in winter, with a white powdery bloom. The catkins, which come in February before the leaves, are striking, the anthers of the male are brick red, turning eventually to yellow.

It will grow accommodatingly in almost any garden soil but is best alongside a stream where the moist conditions encourage long summer growth. It soon makes a twiggy bush and each year's growth is covered with a heavy white and waxy coating of bloom. Allow it to establish itself for a couple of years, then prune it hard in spring to get the best effect for winter.

Handle the stems with care if you pick them for decoration indoors: the white bloom comes off easily and every finger mark will show.
Deciduous shrub   Zone 5-

Salix x rubra 'Eugenei', a hybrid of the purple or bitter osier, is, according to William Robinson, 'a British Willow of some grace of habit', although, at 3m (10ft), not quite a tree.

With its bare branches ascending steeply, it is elegant and conical in shape. Every February, a profusion of slender grey-pink catkins appears, regularly spread along the leafless young and old wood.

If you want to appreciate the purple-hued shoots to the full the following year, then you should coppice the tree fiercely in the spring. The wood of the young shoots is yellow beneath the bark.

This willow can be grown from cuttings, putting half the lengths of 'wand' in the ground where you want your plant to grow.
Deciduous tree   Zones 5-9a

Santolina chamaecyparissus (syn. *S. incana*) is an oral corruption of two Latin words

– *sanctum linum*, literally holy flax. The plant has, however, been known since the 16th century as cotton lavender.

A mass of shining, silvery-white leaves at the very height of winter have long made this aromatic evergreen an ideal choice for low, clipped hedges. Used extensively for parterres and knot gardens, it can equally well be planted as a mound or hummock of grey foliage to make a striking feature for a corner.

Each spring you should cut the shrub back to the old wood – this keeps it in good shape and prevents the vivid yellow button-flowers from putting in an appearance. Reasonably hardy, it should be given plenty of sun and not allowed to get waterlogged during the winter.
Evergreen shrub   Zones 7b-9a

Sarcococca hookeriana digyna, the Christmas box, takes its name from its fruit – *sarkos* being the Greek for flesh and *kokkos* for a berry. It was introduced from west China by E.H. Wilson in 1908 and this species is dedicated to the famous Hookers, father and son, each a director of Kew in his time.

In winter, small white flowers appear first, followed by clusters of fat, black berries so dense that it is the flowers' fragrance rather than their appearance that catches the attention. Although the evergreen foliage resembles box, close inspection reveals that its leaves are alternate, slender and pointed.

Plant this small shrub near a pathway, where you will have the full benefit of the scent. I like to see them set in matching pairs each side of stone steps, where they add architectural interest.
Evergreen shrub   Zones (6-8)

Saxifraga x apiculata, the rockfoil saxifrage, is one of the cushion or Kabschia saxifrages.

This silvery-encrusted plant is among the earliest of the saxifrages to flower and certainly one of the easiest to grow. Its flowers, which appear in spring, are coloured pale lemon and grow in a tight little bunch at the head of single stems, 15cm (6in) high.

This particular saxifrage is quite a hardy survivor and ideal for filling up the crevices in paving, on a terrace or on a path. It is also widely used as a rock garden plant. Like all the saxifrages, it spreads by means of runners with plantlets at the ends.

The Kabschias in particular tend to flower earlier and prefer half-shade. *S.* x 'Kellereri' is one of the first to flower: soft pink sprays above rosettes of grey-green leaves.
Evergreen herbaceous perennial   Zone 6b-

Saxifraga x urbium, London pride, has a rather grandiloquent generic name for such a tiny plant. Taken literally, its name means rock-breaker – *saxum* being the Latin word for rock and *frangere* to break. But this is merely rather an exaggerated reference to its habit of growing in every nook and cranny in its natural mountainous habitat.

In winter some of its foliage turns crimson and forms a weed-smothering mat. In early summer its crop of rich pink 30cm (12in) flowers hover over the carpet of leathery rosettes.

Propagated by division, it grows equally well in sun or partial shade, but if it has a fault it is that it flourishes to a point where it becomes too crowded.
Evergreen perennial groundcover   Zone 7b-

Scilla siberica, the Siberian squill, is one of Linnaeus' generic misappropriations – *scilla* is the Greek vernacular name for the sea squill, a plant which originally he lumped together with this genus but which has now been reclassified.

The deep intense blue of their nodding starry flowers on 12cm (5in) stems is striking. I like to have my scillas coming up through clumps of violets, mixed with chionodoxas and grape hyacinths. They are not ideal for planting in grass but will increase under deciduous trees where little else will flourish in summer.
Bulb   Zone 5-

Sarcococca hookeriana digyna

Saxifraga x *urbium*

Skimmia japonica 'Rubella' was first cultivated in England as early as 1838. The skimmias have been cross bred skilfully to produce male and female clones.

This is the best male clone because its buds are red throughout winter and open in April into ivory-white flowers, with a powerful lily-of-the-valley scent. The best female clone to plant with it is *S.j.* 'Foremanii', or, if you have a small garden, the self-fertile hermaphrodite *S.j. reevesiana* 'Veitchii' from China. Remember that you will get these startling berries only from the female or hermaphrodite clones.

Skimmias like rich, acid soil and prefer partial shade. Most make good groundcover a little under 90cm (36in) high, and many spread by suckering. They are excellent shrubs for seaside or industrial planting.
Evergreen shrub   Zones (7b-8b)

Sorbus aucuparia 'Beissneri' takes its generic name from the fruit of one of its species – the service tree, *S. domestica.*

This 7.5m (25ft) sorbus is a cultivar of the mountain ash. The trunk and stems are a coppery russet, shining and appealing, especially on a wet day. The young shoots complement the trunk with a coral-red glow.

You will have to search for this tree – not many nurseries apart from Hilliers stock it. I first saw it at least 25 years ago in the Cambridge Botanic Gardens and have hankered after it ever since. We ordered and planted one in our 'wilderness', but sadly it turned out to be the wrong species – no russet trunk for which I had kept hoping. The leaves should have told me, since they were oval, not pinnate. Now I have no space for the real 'Beissneri'.

Plant yours where you will come upon it round a corner and remember to stroke the trunk to improve the shine.
Deciduous tree   Zone (2-)

Sorbus vilmorinii commemorates the great French botanist, Maurice de Vilmorin, who sent the first tree to Kew in 1905; it thrives there today. Seeds were collected in China by that remarkable plant collector, the Jesuit priest Père Delavay, and sent to de Vilmorin in 1889.

I consider it one of the best of the mountain ash cultivars for the garden in winter. Its berries last well into December – pale red to begin with, they gradually change to pink and finally become almost white. Loose clusters of white flowers appear in June. A 3.6m (12ft) tree of many seasonal virtues, with fern-like foliage and

rather a spreading habit, it makes an excellent specimen tree where not too much height is required.

Sorbuses are such favourites of mine that we have 14 different varieties of Swedish whitebeam and Mountain ash. They all do well on poor, dry soil and add interest to the garden all the year.
Deciduous tree   Zone 6-

Stachyurus praecox was introduced from Japan in 1864. Its equally hardy cousin, *S. chinensis*, with deeper yellow flowers, was found growing in profusion in China by E.H. Wilson.

From February until April its bare branches drip with what appear to be pendulous strings of tight little primrose bells. If you shake a branch of them you will be in for a surprise: each pendulous shoot is in fact quite stiff and immovable.

It is a shrub of some beauty, with reddish-brown bark and mid-green, slender, pointed leaves which appear after the flowers. It grows well in shade, reaching 3m (10ft) in height, with an equal spread. It likes to be sheltered from the rigours of east and north winds and needs a fairly well-drained soil.
Deciduous shrub   Zones 7b-9a

Stipa gigantea takes its generic name from the Greek word for tow – *stuppe* – the fluffy fibres of hemp or flax used in rope making. This is an allusion to the plant's feathery inflorescences.

Its enormous golden plumes stay on, unbowed, through all weathers, looking marvellous against an evergreen backcloth.

It is best when used on a grand scale so make sure you have the setting right if you wish to incorporate it into your border – it can be overpowering in a small bed. It looks well from a distance and when it is planted with other grasses in a feathery mass of soft, complementary shades.
Grass   Zones 5-9

Sycopsis sinensis was collected by E.H. Wilson on his first foray into China in 1901.

It is a compact shrub or low tree of quiet virtue, not least for its flowers in February and March and for its ability to thrive against a north wall. An upright member of the hamamelis family, it is a free-flowerer, producing clusters of deep yellow blooms or tassels, surrounded by spear-shaped leaves. The yellow stamens are tipped with crimson.

Its natural habitat is high up in the mountains, where it grows as a low tree, but you can see quite a large bush at Kew.
Evergreen shrub   Zone 7b-

Skimmia japonica 'Rubella'

Symphoricarpos orbiculatus 'Foliis Variegatis' announces its fruit-bearing in its generic name. *Symphorein* is the Greek for bear together, *karpos* for fruit – together they reveal that its fruit is carried in clusters. Its common name is the snowberry, and it was introduced into Britain from America as early as 1732.

Its parent plant was a favourite of Thomas Jefferson. He wrote to a friend, 'Its beauty consists in a great produce of berries, literally as white as snow, which remain on the bush through the winter, and make it an object as singular as it is beautiful.'

In winter it has a poise and lightness not often found in winter shrubs – the loss of its leaves, which are prettily marked with a fine outer band of white, reveals a fine and delicate cobweb of branches that contrasts well with the heavier evergreens.
Deciduous shrub   Zones (3b-6b)

Taxus baccata, the common yew, bears unchanged its traditional Latin vernacular name, *taxus*. Native to Britain, it shares with the juniper the distinction of being a coneless conifer.

The yew is beloved of all great English gardeners. Left unclipped, it will achieve a height and spread of 4.5m (15ft). My own experience of this darkest of green-leaved trees is that a 60cm (24in) yew hedge will take 5 years to reach 1.5m (5ft). Spring is the best time to plant yew, summer the best time to water. Clip your new yew in August or September the second year after planting. Its leaves, not its red berries, are poisonous to cattle.
Evergreen tree   Zones 6b-7b

Tellima grandiflora rubra is of particular value to both the winter gardener and flower arranger for the positively purple hue of its leaves.

It is one of those all too rare year-round foliage plants. In summer it bears spikes of pink-fringed, bell-shaped flowers of greenish-white. You can plant it at any time between autumn and spring, increasing your clump by dividing existing plants; it can also be raised from seed.
Evergreen groundcover   Zone (5-)

Teucrium chamaedrys is a Mediterranean evergreen sub-shrub. It is named after King Teucer of Troy, who is said to have recognized its medicinal value.

In late summer and early autumn the spikes of mauvy-pink flowers are attractive, but it is the shining leaves, deep green above and silvery-grey beneath, that make it a welcome if low-growing addition to the garden in winter.

This versatile plant clips well and grows evenly, so you have to decide between letting it flower or clipping it back. I have used it variously – mixed with box and cotton lavender to constitute the threads of my knot garden; as an edging to a low retaining wall facing due north; and as a block of green leaves and autumn flowers in my herb garden, facing due south.
Evergreen sub-shrub   Zones 6a-8a

Thuja plicata 'Atrovirens' is a cultivar of the western red cedar.

Its attraction lies in its bright green foliage and reputation as a fast grower. It will grow up to 15m (50ft) high, but has been a popular choice for evergreen hedges since it was first introduced.

If you use it for a hedge, space the plants 60cm (24in) apart, or stagger them at 90cm (36in) intervals, and you will soon have a firm, solid screen. If you plan to plant them as a specimen group, you should allow 2.75m (9ft) between each to allow them to develop fully.

The natural shape of this thuja is conical, so it should keep itself covered at the base. It clips well, and you should start the first summer after planting, using secateurs or a sharp knife.
Evergreen tree   Zones (4a-7b)

Tiarella wherryi is one species of a genus principally found in North America. The name *tiarella* is an allusion to the way its fruit grows – in the shape of a headband or tiara.

The tiarellas are akin to the heucheras, with their feathery spikes of white or pinkish stellar summer flowers. Gardeners in Britain tend to make more use of *T. cordifolia*, or the foam flower as it is more generally known, for groundcover; in winter its rich green leaves turn a rich shade of bronze. The advantage of *T. wherryi* is that its paler green leaves are attractively flecked with maroon, its racemes are slender and it does not produce stolons.

All the tiarellas grow best in partial shade, propagated by division or grown from seed, they will form low clumps of attractive colour, winter and summer, giving no truck to weeds. They need a well-drained soil, preferably one that is rich in humus.
Evergreen herbaceous perennial   Zone 6

Tilia platyphyllos is the broad-leaved European member of the lime family (linden tree in the USA).

Sycopsis sinensis

Two cultivars which make a noteworthy contribution to the garden in winter are *T.p.* 'Rubra' and *T.p.* 'Aurea'. Both are excellent as 9m (30ft) specimens for the colour of their young twigs and because, unlike the common lime, they are clean-limbed – they do not make adventitious growth on the boles of their trunks. We chose them for our pleached lime walk for this very reason. Later we discovered how colourful and exciting the tops of the trees can look in winter if they are not clipped until early spring.
Deciduous tree   Zones (3b-8a)

Trachycarpus fortunei, the Chusan palm, is native to China, and has been widely grown in Japan for its fibre. Seeds were sent from Japan to the Leiden Botanic Garden in the Netherlands in 1830, and a few years later Robert Fortune collected specimens from the island of Chusan. They grew easily from seed and were first distributed in Britain by Glendinnings nursery in 1860 – a time when exotics were particularly popular. Specimens in the Isle of Wight are now 12m (40ft) high; in the short avenue at Sheffield Park in East Sussex, they are about half that.

It is the only palm which is hardy and which gives an exotic, tropical feel to a garden, but it should be planted in a warm, sheltered place where its large, fan-shaped leaves will be protected from the wind. The yellow male and female flowers, carried on the same plant in May and June, are freely produced and are followed by blue-black fruits.
Evergreen tree   Zones 8b-9a

Tsuga canadensis 'Pendula' is known as Sargent's weeping hemlock.

With dark green leaves white on the undersides, grey-brown bark and green and purple cones, it is one of the best weeping conifers. Broad-headed and pendulous, reaching 1.8m (6ft) in height, it contrasts well with the vertical spires of junipers, better still with the texture of broad-leaved evergreens such as Portugal laurel and *Viburnum tinus*, and makes an excellent foil for hard edges.

Tulipa 'Shakespeare'

There are especially fine examples at Longwood Gardens, Delaware, and I wonder whether any of the 4 original specimens found by General Howland above the Hudson River in about 1857 are still alive. One was certainly given to Charles Sprague Sargent of the Arnold Arboretum.
Evergreen tree   Zones (3a-8a)

Tulipa 'Red Riding Hood' is one of a number of hybrids of *T. greigii*.

The grey-green leaves of the *Greigii* varieties are elegantly marked with purple-brown and bronze marbling and stripes; the flowers are blunt-ended and range from orange to scarlet in colour.

T. 'Red Riding Hood' should be planted on its own, to allow the beauty of its mottled leaves to be enjoyed. These leaves push through in February and lie almost flat on the ground. On warm March days, its scarlet flowers will open wide, on 20cm (8in) stems, and show off their black centres.
Bulb   Zones 4-7

Tulipa kaufmanniana – the waterlily tulip – is a native of Asia, originally from Turkestan.

Unlike most people's idea of a tulip, *T. kaufmanniana* is short-stemmed and sits low on the ground, with scrolls of dark leaves. The first of the tulips, its wide-petalled flowers open in late February. The interior of each flower is creamy-white, yellowing towards the base, and the exterior of the petals throws a red glow over the yellow centre.

Difficult to establish, it is even harder to move, as the bulbs bury themselves so deeply.

There are a number of hybrids derived from the species *T. kaufmanniana*, the flowers of which are mostly in two colour shades. *T.* 'Shakespeare' is red and yellow.
Bulb   Zones 4-8

Viburnum x bodnantense 'Deben' was raised relatively recently by Notcutts Nurseries.

A medium-to-large-sized shrub, it has a strong, upright habit and flowers before shedding its leaves. It draws on the strengths of Lord Aberconway's original hybrid – a cross between *V. farreri* and *V. grandiflorum* – and adds to them *V.f.* 'Candidissimum'. The end result is a viburnum that is neither as gaunt nor as tender as *V. grandiflorum* and is therefore much more akin to the hardy *V. farreri*, but with pure white flowers. *V. x b.* 'Dawn', its better known predecessor, is a vigorous and hardy shrub with masses of frost-resistant flowers between late autumn and winter.

These viburnums are all endowed with a rich, honey scent and are mercifully hardy. To say that every garden should have one is to echo many of the great gardeners. I love them and would not be without them in winter.
Deciduous shrub   Zones (6-8)

Viburnum farreri is W.T. Stearn's renaming of *V. fragrans*. It was first recorded by Pierre d'Incarville, the Jesuit priest, in 1750, growing copiously in north China.

Hard to better, it is a generous 2.75 to 3.6m (9 to 12ft) shrub with pink buds opening to reveal white flowers with an almond scent. It will start to produce its 'apple blossom' flowers in November and go on doing so until March unless it is arrested by a particularly severe frost. There are those who bemoan its reluctance to fruit in cultivation – I will not hear ill of it.
Deciduous shrub   Zones 5a-8a

Viburnum tinus, the common laurustinus, grows wild around the Mediterranean and in south-east Europe. In Britain it has been cultivated since the 16th century and is a favourite with most gardeners.

Its lacy white flowers, which open from pink buds from late autumn through to March, are followed by blue berries that later turn black. My one regret is that of all the viburnums (along with the *V. opulus* varieties), this one should have no scent. *V. tinus* 'Gwenllian' has deeper pink buds and pale pink flowers, and fruits freely.

A hedge of this laurustinus always looks good, but the shrub serves just as well to soften corners. A strong-grower – around 3m (10ft) high and wide – it can be struck from cuttings. It flowers better if its old stems are cut back at intervals.
Evergreen shrub   Zones 8a-9a

Vinca major 'Elegantissima', also known as *V.m.* 'Variegata', is the offspring of the common periwinkle. Its Latin name reminds us of the role periwinkles traditionally played in providing wiry stems for making wreaths – *vincire* means to bind.

V. major 'Elegantissima' is more interesting than *V. minor* – its cream-margined, heart-shaped leaves are larger and of two shades of green, and the plant makes a bigger clump.

Although not universally popular among gardeners, vincas can provide essential groundcover. We clip ours back hard during winter, allowing new shoots to develop and flower in spring. The exception is *V. difformis*,

whose blue, interestingly shaped flowers appear in winter and can be clipped later.
Herbaceous perennial   Zone (7b-)

Viola riviniana, one of the hardiest plants for groundcover, is the purple-leaved form of the European dog violet.

It has purplish leaves which spread quickly into mats of welcome colour for winter, and light purple-blue flowers in spring.

I have seen it used successfully to soften the edges of borders or to fill the crevices of paving. I use mine in terracotta pots to provide winter colour at the base of clipped variegated box. This viola seeds itself and, once established, you will have plenty of plants to give away.
Herbaceous perennial   Zones 4-8

Viscum album, as its name fully suggests, has sticky white berries – *viscus* being the Latin word for stickly and *albus* for white. Better known as mistletoe, it is native to Europe, North Africa and northern Asia. In the USA it has an altogether different botanical name – *Serotinum phoradendron.*

A colourful parasite, the whole plant hangs down from the branches of its 'host' tree in a short cluster, 30 to 60cm (12 to 24in) long. Its thick, pliable stems and leaves are yellowish-green, its one-seed berries translucent. Although tradition has it living off the oak tree, it is more likely to be found on old apple trees, thorns, ashes, hazels and maples.

When birds eat these berries, the sticky flesh adheres to their beaks. When they clean their beaks against the tree bark, you have mistletoe in the making. Thus, to propagate it, rub ripe seed into the bark of the tree you wish to colonize.
Evergreen shrub   Zone 7

Waldsteinia ternata is to be found throughout central Europe and as far east as Siberia and Japan.

This low-growing evergreen produces sheaves of bright yellow flowers in late spring or early summer and provides excellent groundcover in winter. Its rich, green, trifoliate leaves, lying in great mats, look much like those of alpine strawberry plants. The leaflets are irregularly toothed and ovate in shape.

It is one of those rare groundcoverers that grows in shade or sun. In shade it will establish and increase more rapidly; in sun it will flower more readily, provided the soil does not dry out. Once established, the mats are easily split, but they must be replanted immediately.
Herbaceous perennial   Zone 4-

Viburnum opulus

PLANT ZONING

In the preceding plant portraits hardiness zone ratings are given to indicate the approximate minimum temperature a plant will tolerate in winter, following the system used by the United States Department of Agriculture. Each zone corresponds to a range of average minimum winter temperatures.

Plants are given two zone ratings, the first indicating the northernmost zone in which the plant is likely to survive and the second its southernmost limit. Where no second zone is given, this is because specific information is lacking.

All zones are further subdivided into north and south, indicated by 'a' and 'b' respectively, to cater for plants which may be hardy only in the northern or southern regions of a particular zone.

Parantheses are used where zoning given is for the species rather than the cultivar which, although bound to thrive in similar conditions, may be more or less hardy.

A plant's hardiness can vary within quite a small area depending on the altitude, exposure, moisture and type of soil of a particular site.

The chart below defines the minimum temperature range for each zone:
1   below $-45°C/-50°F$
2   $-45°C/-50°F$ to $-40°C/-40°F$
3   $-40°C/-40°F$ to $-34°C/-30°F$
4   $-34°C/-30°F$ to $-29°C/-20°F$
5   $-29°C/-20°F$ to $-23°C/-10°F$
6   $-23°C/-10°F$ to $-18°C/0°F$
7   $-18°C/0°F$ to $-12°C/10°F$
8   $-12°C/10°F$ to $-7°C/20°F$
9   $-7°C/20°F$ to $-1°C/30°F$
10 $-1°C/30°F$ to $4°C/40°F$

162

Bibliography

ACTON, Sir Harold *Villas in Tuscany* (Thames & Hudson, London, 1973)

ALLAN, M.E.A.. *E.A. Bowles and his garden at Myddleton 1865-1954* (Faber, London, 1973)

BACON, Francis Essay 'Of Gardens' in *Bacon's Essays* (T. Nelson & Sons)

BATSON, Wade T. *Landscape Plants for the Southeast* (University of South Carolina, Columbus, 1984)

BECKETT, Kenneth A. *The Concise Encyclopedia of Garden Plants* (Orbis, London, 1978)

BOWLES, E.A. *My Garden in Autumn & Winter* (T.C. & E.C. Jack, London, 1915)

BOWLES, E.A. *My Garden in Spring* (T.C. & E.C. Jack, London, 1915)

BRUCE, Harold *Winterthur in Bloom* (Merrill, Columbus, Ohio, 1968; Winterthur, 1986, paperback edition)

CHATTO, Beth *The Dry Garden* (Dent, London, 1983)

CHURCH, Thomas D. and HALL, Grace and LAURIE, Michael *Gardens are for People* (McGraw Hill, New York, 1976, 2nd edition)

COATS, Alice M. *The Quest for Plants* (Studio Vista,London, 1969; McGraw Hill, New York, 1970, as *The Plant Hunters*)

DUTTON, Joan Parry *Plants of Colonial Williamsburg* (Colonial Williamsburg Foundation, Virginia, 1979)

EARLE, C.W. and E.V.B. and KINGSLEY, Rose and GIBBS, Hon. Vicary *Garden Colour* (J.M. Dent, London, 1905)

ELLACOMBE, Henry, Canon *In a Gloucestershire Garden* (Arnold, London, 1895; Century Publishing, London, 1982)

EMBERTON, Sybil *A Year in the Shrub Garden* (Faber, London, 1972)

FISH, Margery *An All the Year Garden* (Collingridge, London, 1964)

FISH, Margery *Gardening in the Shade* (Collingridge, London, 1964; Faber & Faber, 1983, paperback edition)

FISH, Margery *Ground Cover Plants* (Collingridge, London, 1964; 3rd impression, 1965; Faber & Faber, 1980, paperback edition)

FISH, Margery *We Made a Garden* (Collingridge, London, 1956; 4th impression, Newton Abbott, 1970; Faber & Faber, 1983, paperback edition)

FREDERICK, William *100 Great Garden Plants* (Alfred Knopf, New York, 1975)

GARDEN CLUB OF AMERICA (Janet M. Poor ed.) *Plants that Merit Attention*, Volume I, Trees (Timber Press, Portland, Oregon, 1984)

GIBSON, John and WEATHERLY, Neal *The Gardens of Williamsburg* (Carlisle H. Humelsine ed.) (A Colonial Williamsburg Publication, 1970)

GIBSON, John and WEATHERLY, Neal *Trees for the Landscape, Selection and Culture* (University of Georgia, 1982)

GOMBRICH, E.H. *The Sense of Order – A Study in the Psychology of Decorative Art* (Phaidon Press, Oxford, 1984, 2nd edition)

GRIGSON, Geoffrey *The Englishman's Flora* (Phoenix House, 1958)

HARPER, Pamela and McGOURTY, Frederick *Perennials, How to Select, Grow and Enjoy* (H.P. Books, Tucson , Arizona, 1985)

HIBBERD, Shirley *The Ivy, its History, Uses and Characteristics* (Groombridge, London, 1872)

HICKS, David *Garden Design* (Routledge & Kegan Paul, 1982)

HILL, Thomas *The Proffitable Arte of Gardening* (1568)

HILLIER NURSERIES *The Hillier Colour Dictionary of Trees and Shrubs* (David & Charles, Newton Abbott, 1981)

HOBHOUSE, Penelope *Colour in Your Garden* (Collins, London, 1988; Little Brown, Boston, 1985)

HOBHOUSE, Penelope *Gertrude Jekyll on Gardening* (Collins, London, 1983)

HUNT, William Lanier *Southern Gardens, Southern Gardening* (Duke, Durham, North Carolina, 1982)

JEKYLL, Gertrude *Wood and Garden* (Longmans & Co., London, 1899)

JOHNSON, Hugh and MILES, Paul *The Mitchell Beazley Pocket Guide to Garden Plants* (Mitchell Beazley, London, 1981)

KELLAM DE FOREST, Elizabeth *The Gardens and Grounds at Mount Vernon* (M.U.L.A., Mount Vernon, Virginia, 1982)

LAWRENCE, Elizabeth *Gardens in Winter* (Harper, New York, 1961)

LAWSON, William *The Country House-Wife's Garden* (1617)

LAY, Charles Downing *A Garden Book for Autumn and Winter* (Duffield, New York, 1924)

LEIGHTON, Ann *American Gardens in the Eighteenth Century 'For Use and For Delight'* (Houghton Mifflin Co., Boston, 1976; University of Massachusetts Press, 1986, paperback edition)

LLOYD, Christopher *Foliage Plants* (Collins, London, 1975; Random House, New York, 1985)

LLOYD, Christopher *The Well-Tempered Garden* (Collins, London, 1970; Random House, New York, 1985)

MARKHAM, Gervase *Maison Rustique, OR The Country Farme* (1616)

MATHEW, Brian *Dwarf Bulbs* (Batsford, London, 1978)

MOUNTAINE, Didymus *The Gardeners Labyrinth*

NEWHALL, Charles *The Shrubs of Northeastern America* (Putnam, New York, 1897)

PAGE, Russell *The Education of a Gardener* (Collins, London, 1962; Random House, New York, 1985)

PERENYI, Eleanor *Green Thoughts – A Writer in the Garden* (Random House, New York, 1981; paperback edition, 1983)

RAKUSEN, Philippa *Foliage and Form throughout the Year* (ill. Barbara N. Shaw) (Harlow Car Enterprises)

ROBINSON, William *The English Flower Garden and Home Grounds* (John Murray, London, 1906, 10th edition)

ROYAL HORTICULTURAL SOCIETY *Dictionary of Gardening* (Clarendon Press, Oxford, 1956, 2nd edition)

SEDGE, Mable Cabot *The Garden Month by Month*

SUTTON, S.B. *The Arnold Arboretum – The First Century* (Random House, New York, 1981)

SYNGE, Patrick *Flowers & Colour in Winter* (Lindsay Drummond, 1940; Michael Joseph, 1974)

THOMAS, Graham Stuart *Colour in the Winter Garden* (Dent, London, 1957; 3rd edition, 1984)

Trustees' Garden Club, Savannah, Georgia, *Garden Guide to the Lower South* (Moran Printing Company, Florida, 1986)

WHITE, Gilbert *Garden Kalendar* 1751-1771 (Scolar Press, London, 1975)

WILSON, E.H. *America's Greatest Garden* (Stratford, Boston, 1925)

WOOD, Louisa Ferrand and ELLIS, Ray *Behind those Garden Walls in Historic Savannah* (Historic Savannah Foundation, Savannah, Georgia, 1982)

Index

Author's Acknowledgments

When first I embarked on this book, I thought I would write to some of my gardening friends and to other gardeners I admired, to ask them how they felt about their garden in winter. Their letters were so illuminating and in many cases so reflective of my own thoughts that I decided to weave some of them into the fabric of my book. I am grateful to them all, especially Lord Aberconway, Hardy Amies, Dick Balfour, Bob Dash, Valerie Finnis, William Frederick, Ryan Gainey, Penelope Hobhouse, Hugh Johnson, Christopher Lloyd, Allen Paterson, Pippa Rakusen, Nicholas Ridley, John Sales, Marina Schinz, Reresby Sitwell, Roy Strong and James van Sweden.

Then there are those people who kept me up to the mark with their encouragement, advice and hospitality. Their visits, letters and telephone calls helped me over many an attack of writer's block and typist's cramp. Among them are Tom Woodham, Marilyn Alaimo, Michael Balston, John Cook, Dagny Holland-Martin and Peter Coats.

I must also thank those of my gardening friends who also find time to be writers for allowing me to quote from their published works: Beth Chatto, William Frederick, David Hicks, Christopher Lloyd, Anne Scott-James and Graham Stuart Thomas.

My thanks too to the publishers of their books and those of other authors from whose works I have quoted: Collingridge, Collins, Country Life, J.M. Dent, Harper, Michael Joseph, Alfred Knopf, Merrill, Routledge & Kegan Paul, Thames & Hudson.

My special thanks to Tim Rock and Katherine Lambert whose idea it was that I should write this book and for the many hours they spent helping me; to Andrew Lawson for his superb photographs; to the team, including Erica Hunningher, Anne Wilson, Anne Fraser, Gian Douglas Home and Barbara Vesey, at Frances Lincoln Ltd., who are responsible for the final publication; to Paul Miles, Tim Rees and Sue Spielberg for helping me with research; to Tony Lord of the National Trust for guiding me through the minefield of plant names, and for compiling the index; and to Rick Darke, curator of plants at Longwood Gardens, Pennsylvania, and Robert Herald, curatorial assistant, for categorizing the plants according to their US hardiness zones. John Nelly and Marco Polo Stufano of Wave Hill Gardens, the Bronx, New York patiently answered all my questions about East Coast American plants.

Publishers' Acknowledgments

The publishers would like to thank Media Typesetting & Productions Ltd. for typesetting and Universal Colour Scanning Ltd., Hong Kong, for colour reproduction.

They also gratefully acknowledge permission to quote from the following publications:
(pp. 56 & 91) *Colour in the Winter Garden* Graham Stuart Thomas (J. M. Dent & Sons Ltd., London)
(p. 76) *Garden Design* David Hicks (Routledge & Kegan Paul Ltd., London)
(p. 86) *100 Great Garden Plants* © 1975 William Frederick (Alfred A. Knopf, Inc., New York)
(p. 88) *Villas in Tuscany* © 1973 Sir Harold Acton (Thames & Hudson Ltd., London)
(p. 124) *Winterthur in Bloom* Harold Bruce (Winterthur Museum & Gardens, Delaware)

Photographic Credits

All photography by Andrew Lawson except the following:
(L=left R=right T=top B=bottom M=middle O=owned by)
Heather Angel/Biofotos: 1, 64, 132L, 147L
A-Z Collection: 70, 133L, 150R, 152R, 153, 154L
Jill Bailey: 81
Peter Baistow: 54, 82
R.C. Balfour: Back cover, 20, 73, 103, 132R, 134L, 137, 139B, 143 149M, 155
Geoff Dann: 142L © FLL
Derek Fell: 6, 40, 60, 108, 113, 114
John Fielding: 107, 154M
Valerie Finnis: 125
Jerry Harpur: 80
Marijke Heuff: 14 (O: Canneman), 27 (O: De Wildenborch), 39 (O: Nieuwenhuis), 52/3 (O: van Bennekom)
Jacqui Hurst: 71, 72
Nada Jennett: 51, 66/67
Georges Lévêque: 74, 79, 87, 105
Tony Lord: 29, 32, 78, 126R
S & O Mathews: 16, 111, 142M, 147M, 160
Paul Miles: 133R
Nature Photographers Ltd (Frank V Blackburn): 123
Oxford Scientific Films Ltd (G A Maclean): 122
Tim Rees: 121
Richard Robinson: 34
Harry Smith Horticultural Photographic Collection: 139T, 154R
M.P. Stufano: 145, 151R
Jessica Strang: 30
Pamla Toler/IMPACT Photos: 77
Rosemary Verey: 89
Michael Warren/Photos Horticultural: 95, 126L, 130, 134R, 135, 136, 140, 141, 144T, 146, 148, 149T, 149B, 150L, 151L, 152L, 157R, 158, 159